THE QUICK REFERENCE WORLD ATLAS

This edition produced 1995 by The Promotional Reprint Company Ltd,
Deacon House, 65 Old Church Street,
Chelsea, London, SW3 5BS

Published in the USA 1995 by JG Press
Distributed by World Publication, Inc.

The JG Press imprint is a trademark of JG Press, Inc.,
455 Somerset Avenue, North Dighton, MA 02764

Computer Cartography designed and produced by European Map Graphics Ltd.
Alberto House, Hogwood Lane,
Finchampstead, Berkshire.

ISBN 1 57215 093 9

Flags produced by E.M.G. Ltd and authenticated by The Flag Research Center,
Winchester, Mass. 01890 U.S.

Printed and bound in China

Contents

World

World - Key Map 4-5
World - Physical 6-7
World - Ocean Currents 8
World - Earthquake Zones 8
World - Global Warming 9
World - Acid Rain 9
Flags of the Worlds Nations 10-13

Europe

Europe - Political 14-15
Albania, Andorra, Austria 16
Belgium, Belarus, Balearic Islands 17
Bosnia Herzegovina, Bulgaria, Croatia 18
Cyprus, Czech Republic, Denmark 19
Estonia, France, Monaco 20
Finland, Germany 21
Greece, Hungary, Iceland 22
Ireland, Italy, San Marino, Vatican City 23
Latvia, Liechtenstein, Lithuania, Luxembourg 24
Macedonia, Malta, Moldova 25
Netherlands, Norway, Poland 26
Portugal, Romania, Slovak Republic 27
Slovenia, Spain, Sweden 28
Switzerland, Turkey, Ukraine, Yugoslavia 29
United Kingdom 30-31

North and Central America

North &Central America -Political 32-33
Antigua & Barbuda, Bahamas, Barbados 34
Belize, Bermuda, Costa Rica 35
Canada, Cuba, Dominica 36
Dominican Republic, El Salvador, Greenland 37
Grenada, Guatemala, Haiti, Honduras 38
Jamaica, Mexico, Nicaragua, Panama 39
Puerto Rico, St. Christopher Nevis, St Lucia, St Vincent 40
Trinidad & Tobago, United States 41

South America

South America - Political 42-43
Argentina, Bolivia, Brazil 44
Chile, Colombia, Ecuador 45
Guyana, Paraguay, Peru 46
Surinam, Uruguay, Venezuela 47

Africa

Africa - Political 48-49
Algeria, Angola, Azores 50
Benin, Botswana, Burkina Faso 51
Burundi, Cameroon, Canary Islands 52
Cape Verde, Central African Republic, Chad 53
Comoros, Congo, Cote d'Ivoire, Djibouti 54
Egypt, Eritrea, Ethiopia, Equatorial Guinea 55
Gabon, Gambia, Ghana 56
Guinea, Guinea Bissau, Kenya 57
Lesotho, Liberia, Libya 58
Madagascar, Madeira, Malawi 59
Mali, Mauritania, Mauritius 60
Morocco, Mozambique, Namibia 61
Niger, Nigeria, Rwanda 62
São Tomé & Príncipe, Senegal, Seychelles, Sierra Leone 63
Somalia, South Africa, Sudan 64
Swaziland, Tanzania, Togo 65
Tunisia, Uganda, Zaire 66
Zambia, Zimbabwe 67

Asia

Asia - Political 68-69
Afghanistan, Armenia, Azerbaijan 70
Bahrain, Bangladesh, Bhutan 71
Brunei, Cambodia, Georgia 72
China, Hong Kong 73
India, Indonesia, Iran 74
Iraq, Israel, Japan 75
Jordan, North Korea, South Korea 76
Kazakhstan, Kuwait, Kyrgyzstan 77
Laos, Lebanon, Malaysia 78
Maldives, Mongolia, Myanmar(Burma) 79
Nepal, Oman, Pakistan 80
Philippines, Qatar 81
Russia, Saudi Arabia 82
Singapore, Sri Lanka, Syria 83
Taiwan, Tajikistan, Thailand, Turkmenistan 84
United Arab Emirates, Uzbekistan, Vietnam, Yemen 85

Oceania and Polar Regions

Oceania - Political 86-87
Australia, Fiji, Kiribati & Tuvalu 88
New Zealand, Northern Marianas, Nauru, Marshall Islands 89
Papua New Guinea, Palau, Soloman Islands 90
Tonga, Vanuatu, Western Samoa 91
Arctic & Antarctica 92

Index to Placenames 93-96

Political Key Map

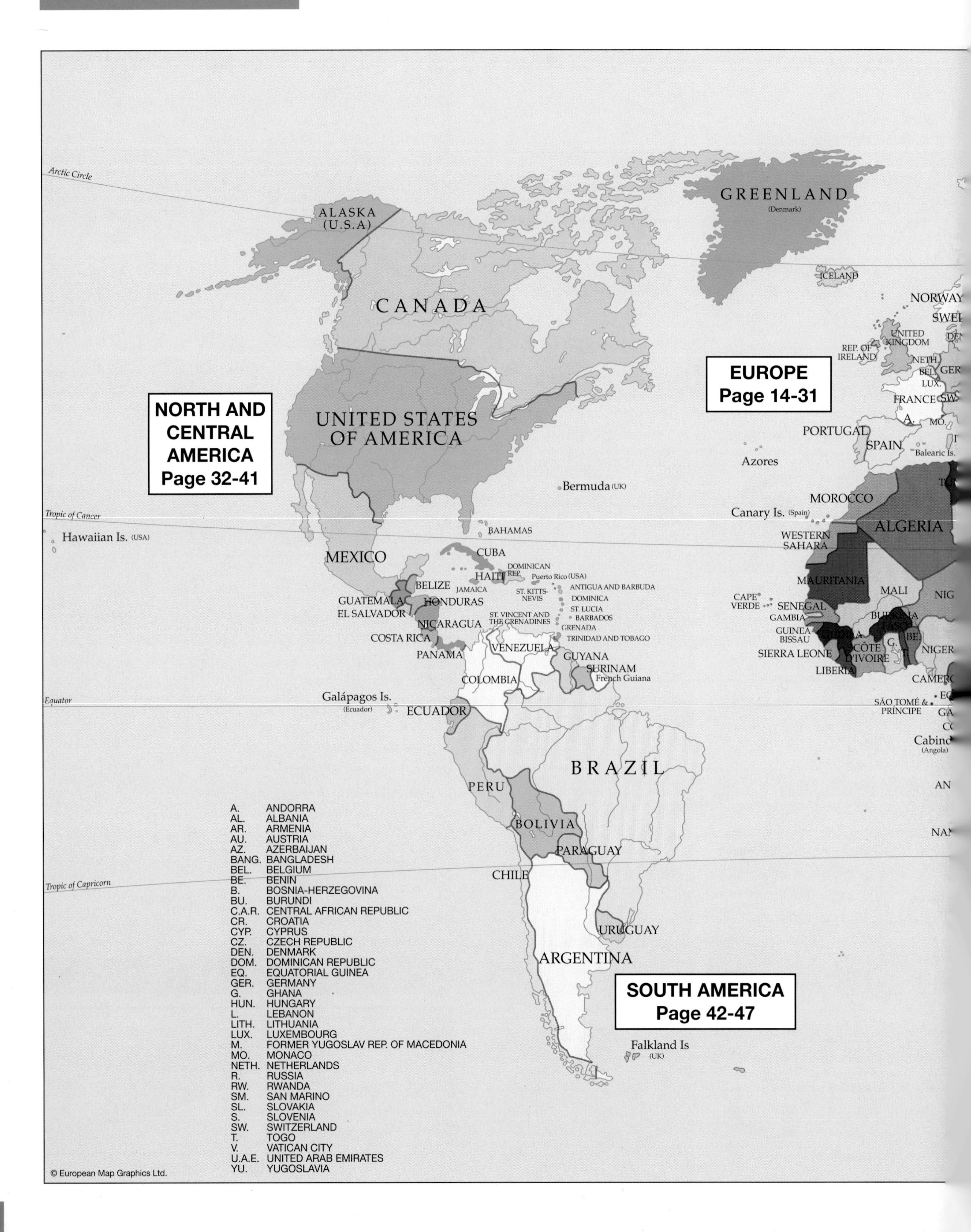

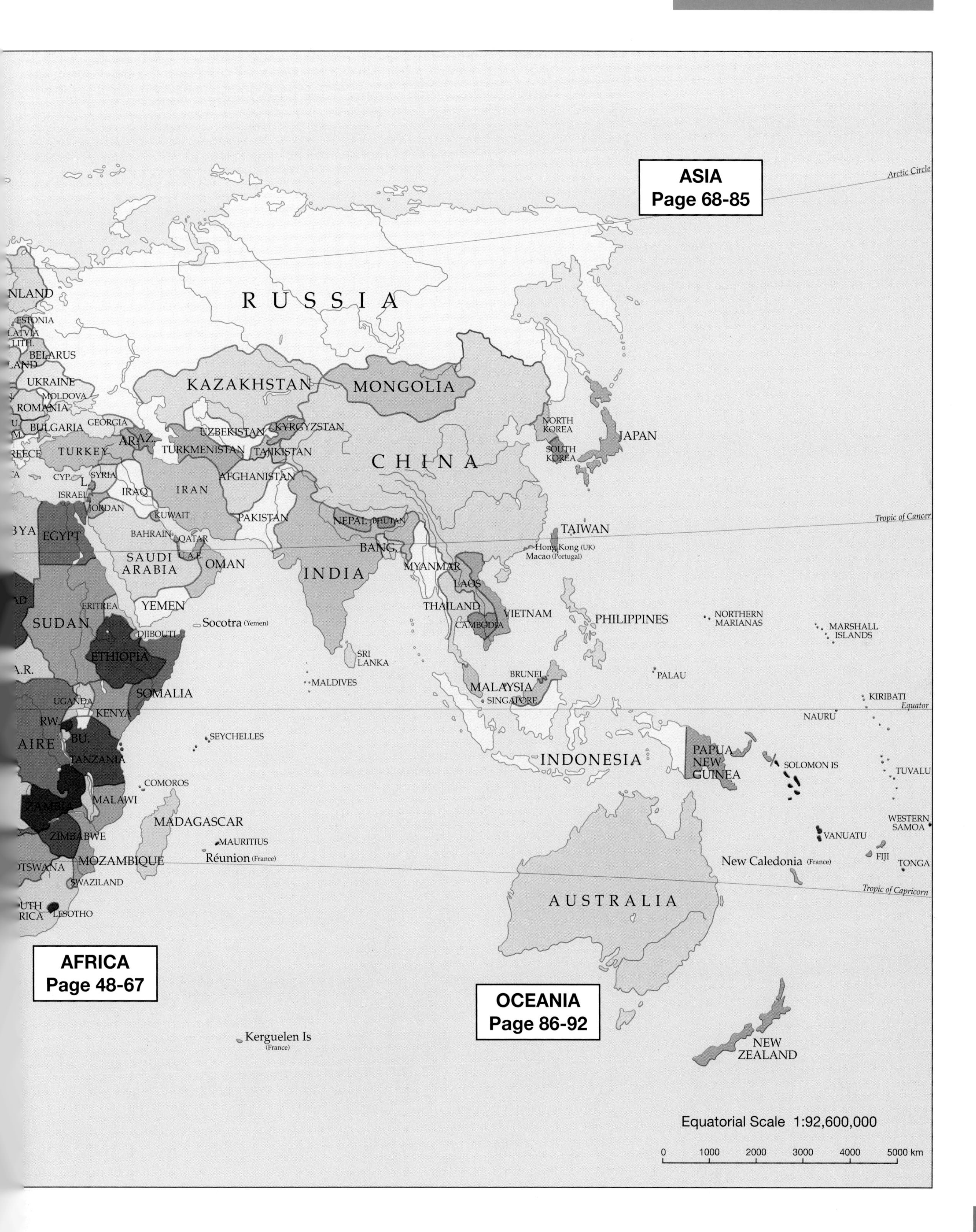
ASIA
Page 68-85
Arctic Circle
RUSSIA
KAZAKHSTAN
MONGOLIA
UZBEKISTAN
KYRGYZSTAN
TURKMENISTAN
TAJIKISTAN
NORTH KOREA
SOUTH KOREA
JAPAN
CHINA
AFGHANISTAN
IRAN
IRAQ
PAKISTAN
NEPAL
BHUTAN
TAIWAN
Tropic of Cancer
Hong Kong (UK)
Macao (Portugal)
BANG.
INDIA
MYANMAR
LAOS
THAILAND
VIETNAM
CAMBODIA
PHILIPPINES
NORTHERN MARIANAS
MARSHALL ISLANDS
SRI LANKA
MALDIVES
BRUNEI
MALAYSIA
SINGAPORE
PALAU
KIRIBATI
Equator
NAURU
INDONESIA
PAPUA NEW GUINEA
SOLOMON IS
TUVALU
WESTERN SAMOA
VANUATU
FIJI
TONGA
New Caledonia (France)
Tropic of Capricorn
AUSTRALIA
OCEANIA
Page 86-92
NEW ZEALAND
NLAND
ESTONIA
LATVIA
LITH.
BELARUS
UKRAINE
MOLDOVA
ROMANIA
BULGARIA
GEORGIA
AR.
AZ.
TURKEY
CYP.
SYRIA
L.
ISRAEL
JORDAN
KUWAIT
BAHRAIN
QATAR
U.A.E.
SAUDI ARABIA
OMAN
YEMEN
Socotra (Yemen)
EGYPT
ERITREA
SUDAN
DJIBOUTI
ETHIOPIA
SOMALIA
UGANDA
KENYA
RW.
BU.
AIRE
TANZANIA
SEYCHELLES
COMOROS
MALAWI
ZAMBIA
ZIMBABWE
MADAGASCAR
MAURITIUS
Réunion (France)
MOZAMBIQUE
SWAZILAND
LESOTHO
AFRICA
Page 48-67
Kerguelen Is (France)
Equatorial Scale 1:92,600,000
0 1000 2000 3000 4000 5000 km

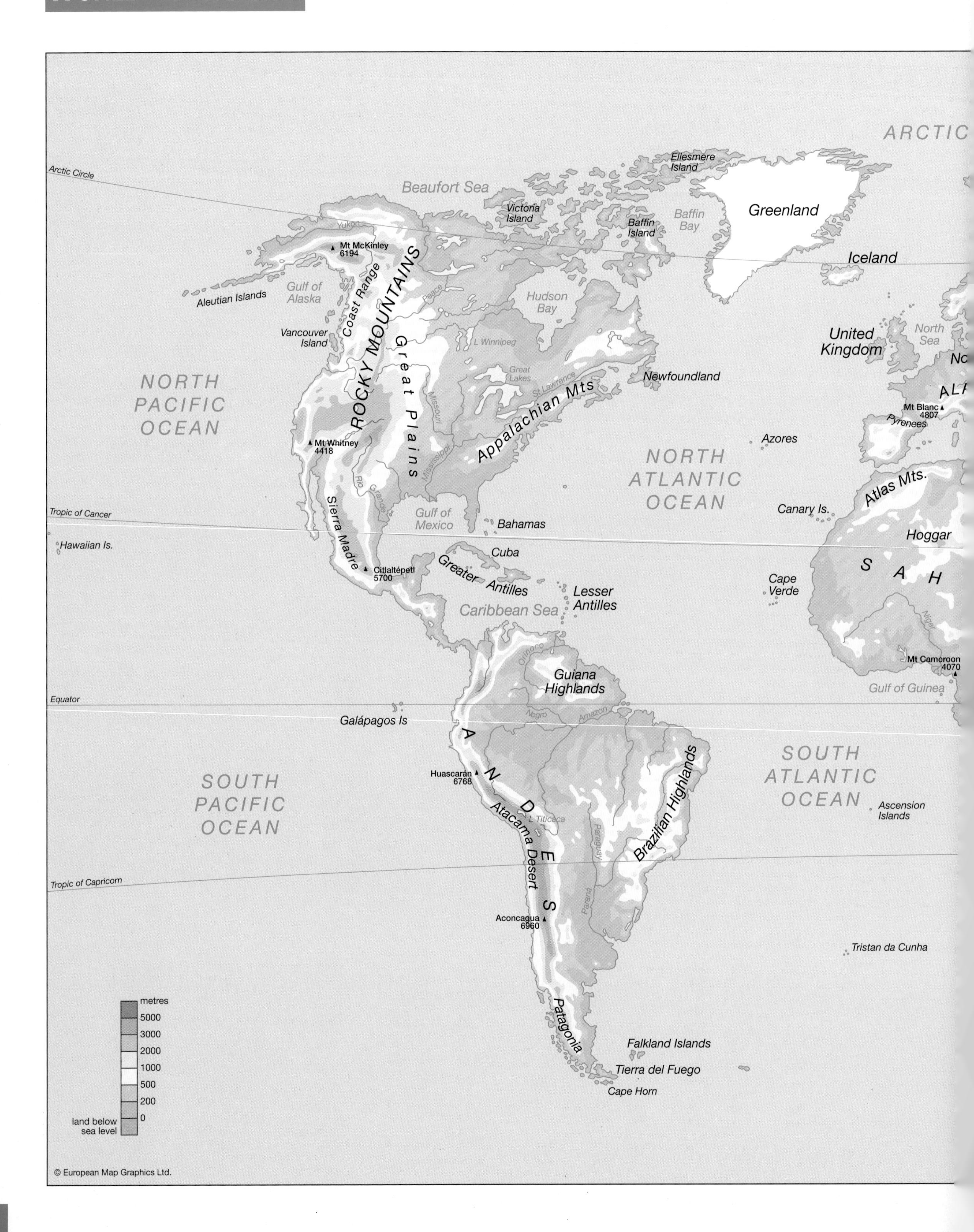
ARCTIC
Arctic Circle
Ellesmere Island
Beaufort Sea
Victoria Island
Greenland
Baffin Island
Baffin Bay
Yukon
Mt McKinley 6194
Iceland
Aleutian Islands
Gulf of Alaska
Coast Range
ROCKY MOUNTAINS
Peace
Hudson Bay
Vancouver Island
Great Plains
L Winnipeg
United Kingdom
North Sea
Great Lakes
St Lawrence
Newfoundland
NORTH PACIFIC OCEAN
Appalachian Mts
Missouri
Mt Blanc 4807
Pyrenees
Mt Whitney 4418
Mississippi
Azores
NORTH ATLANTIC OCEAN
Rio Grande
Atlas Mts.
Sierra Madre
Canary Is.
Tropic of Cancer
Gulf of Mexico
Bahamas
Hoggar
Hawaiian Is.
Cuba
Citlaltépetl 5700
Greater Antilles
Lesser Antilles
Cape Verde
Caribbean Sea
Niger
Orinoco
Mt Cameroon 4070
Guiana Highlands
Gulf of Guinea
Equator
Negro
Amazon
Galápagos Is
SOUTH ATLANTIC OCEAN
Huascarán 6768
Brazilian Highlands
SOUTH PACIFIC OCEAN
Ascension Islands
ANDES
Atacama Desert
L Titicaca
Paraguay
Tropic of Capricorn
Paraná
Aconcagua 6960
Tristan da Cunha
metres
5000
3000
2000
1000
500
200
0
land below sea level
Patagonia
Falkland Islands
Tierra del Fuego
Cape Horn
© European Map Graphics Ltd.

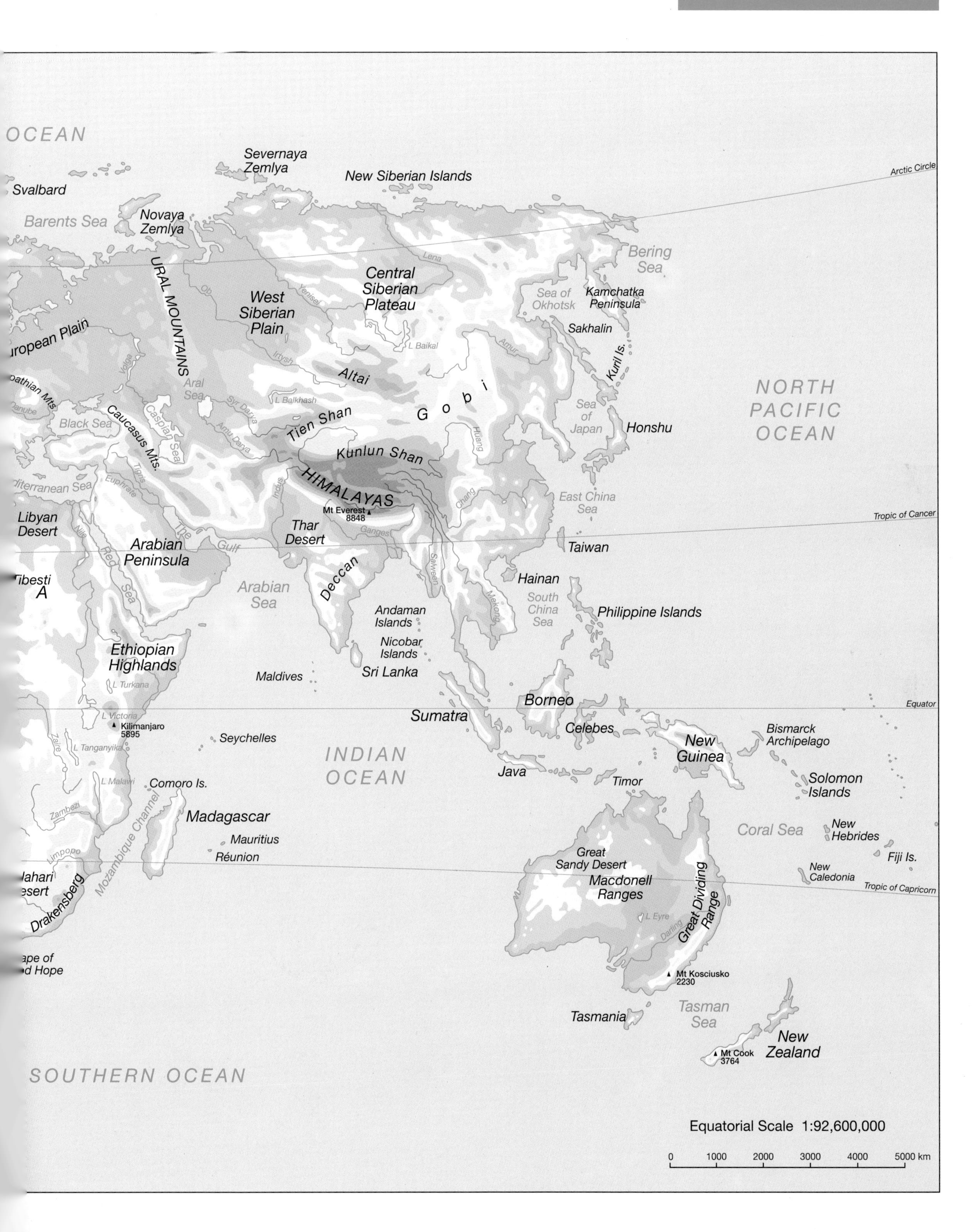
OCEAN
Severnaya Zemlya
New Siberian Islands
Arctic Circle
Svalbard
Barents Sea
Novaya Zemlya
URAL MOUNTAINS
Lena
Central Siberian Plateau
Bering Sea
Ob
Yenisei
West Siberian Plain
Sea of Okhotsk
Kamchatka Peninsula
uropean Plain
Sakhalin
L Baikal
Amur
Volga
Irtysh
Kuril Is.
Altai
pathian Mts
Aral Sea
NORTH PACIFIC OCEAN
Danube
L Balkhash
Gobi
Sea of Japan
Caucasus Mts.
Caspian Sea
Syr Darya
Black Sea
Amu Darya
Tien Shan
Honshu
Huang
Kunlun Shan
Tigris
diterranean Sea
Euphrates
Indus
HIMALAYAS
East China Sea
Chang
Libyan Desert
Mt Everest 8848
Tropic of Cancer
Nile
The Gulf
Thar Desert
Ganges
Arabian Peninsula
Taiwan
Red Sea
Deccan
Salween
Tibesti
Hainan
Arabian Sea
South China Sea
Mekong
Andaman Islands
Philippine Islands
Nicobar Islands
Ethiopian Highlands
Sri Lanka
Maldives
L Turkana
Borneo
Equator
L Victoria
Sumatra
Kilimanjaro 5895
Celebes
Bismarck Archipelago
Seychelles
Zaire
L Tanganyika
New Guinea
INDIAN OCEAN
Java
Timor
Solomon Islands
L Malawi
Comoro Is.
Zambezi
Madagascar
Coral Sea
New Hebrides
Mozambique Channel
Mauritius
Great Sandy Desert
Fiji Is.
Limpopo
Réunion
New Caledonia
lahari esert
Macdonell Ranges
Great Dividing Range
Tropic of Capricorn
Drakensberg
L Eyre
Darling
ape of d Hope
Mt Kosciusko 2230
Tasmania
Tasman Sea
New Zealand
Mt Cook 3764
SOUTHERN OCEAN
Equatorial Scale 1:92,600,000
0
1000
2000
3000
4000
5000 km

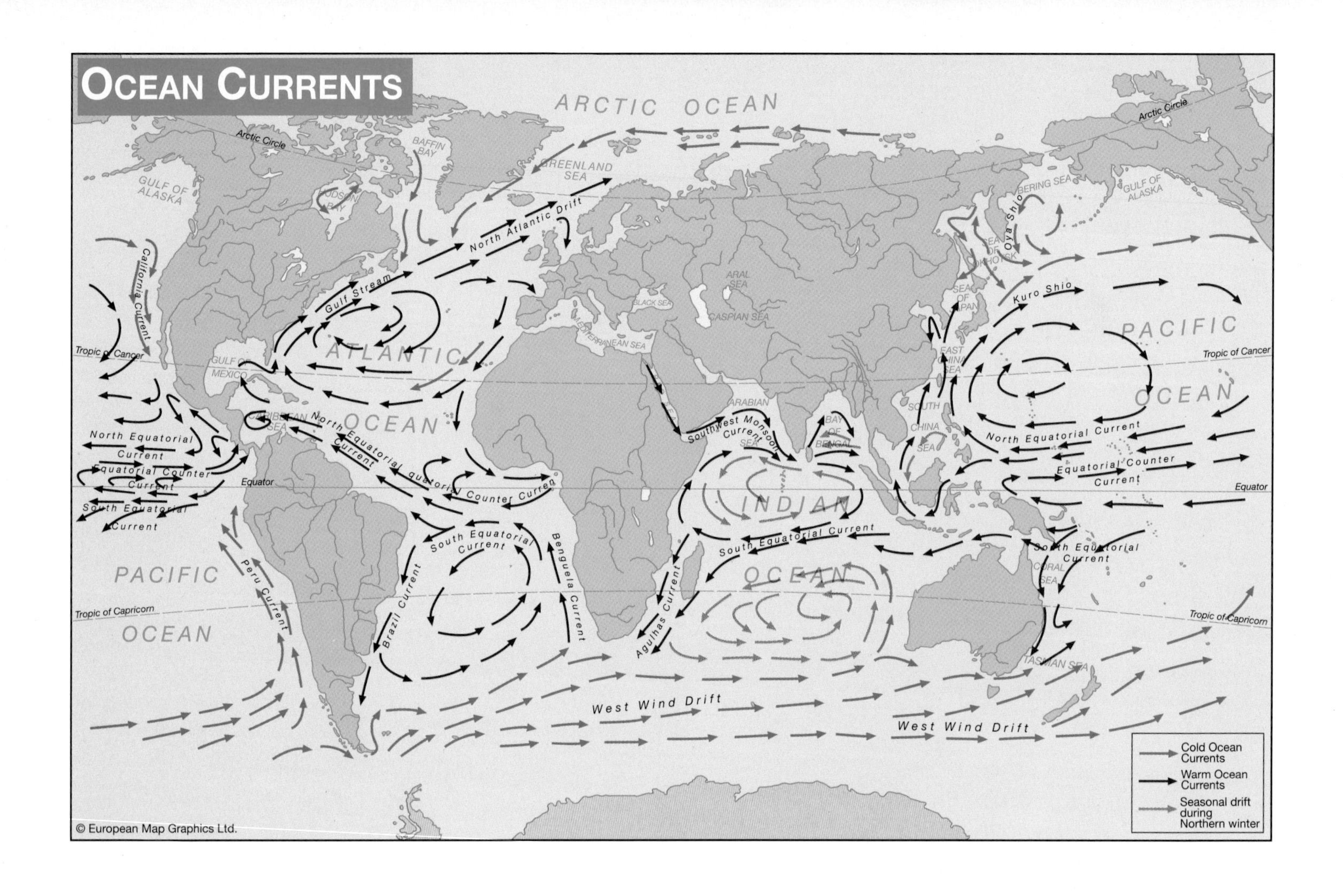

OCEAN CURRENTS
ARCTIC OCEAN
Arctic Circle
BAFFIN BAY
GREENLAND SEA
GULF OF ALASKA
HUDSON BAY
North Atlantic Drift
California Current
Gulf Stream
ATLANTIC
OCEAN
Tropic of Cancer
GULF OF MEXICO
CARIBBEAN SEA
North Equatorial Current
Equatorial Counter Current
South Equatorial Current
Equator
Equatorial Counter Current
South Equatorial Current
Benguela Current
Brazil Current
PACIFIC
OCEAN
Peru Current
Tropic of Capricorn
West Wind Drift
ARAL SEA
BLACK SEA
CASPIAN SEA
MEDITERRANEAN SEA
ARABIAN SEA
Southwest Monsoon Current
BAY OF BENGAL
INDIAN
OCEAN
Agulhas Current
BERING SEA
Oya Shio
SEA OF OKHOTSK
SEA OF JAPAN
Kuro Shio
EAST CHINA SEA
SOUTH CHINA SEA
PACIFIC
OCEAN
North Equatorial Current
Equatorial Counter Current
South Equatorial Current
CORAL SEA
TASMAN SEA
Cold Ocean Currents
Warm Ocean Currents
Seasonal drift during Northern winter
© European Map Graphics Ltd.

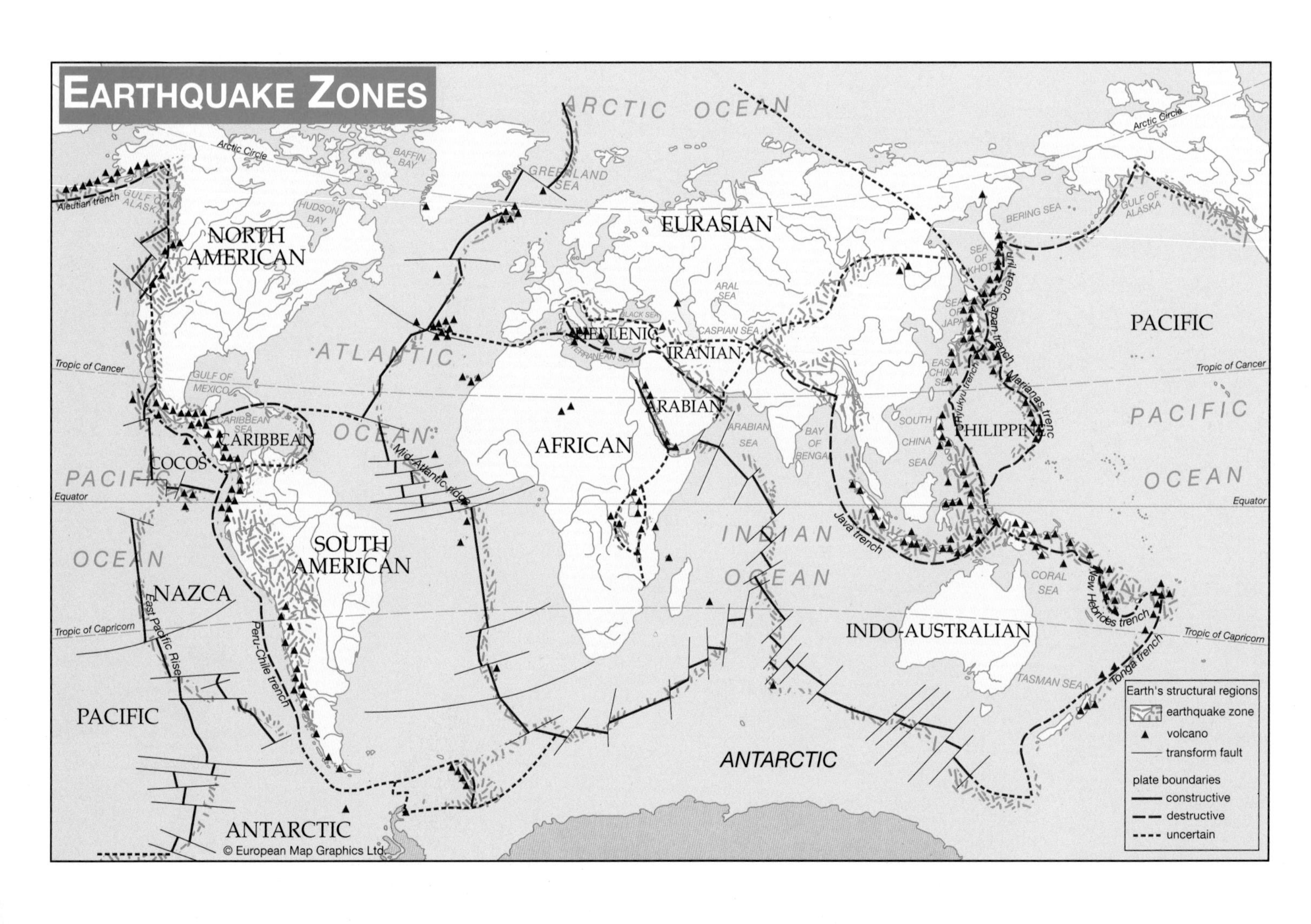

EARTHQUAKE ZONES
ARCTIC OCEAN
Arctic Circle
Aleutian trench
GULF OF ALASKA
BAFFIN BAY
GREENLAND SEA
HUDSON BAY
NORTH AMERICAN
EURASIAN
ARAL SEA
BLACK SEA
CASPIAN SEA
HELLENIC
IRANIAN
ATLANTIC
OCEAN
Tropic of Cancer
GULF OF MEXICO
CARIBBEAN
ARABIAN
ARABIAN SEA
BAY OF BENGAL
AFRICAN
COCOS
Mid-Atlantic ridge
PACIFIC
Equator
OCEAN
SOUTH AMERICAN
NAZCA
East Pacific Rise
Peru-Chile trench
Tropic of Capricorn
PACIFIC
INDIAN
OCEAN
Java trench
INDO-AUSTRALIAN
ANTARCTIC
ANTARCTIC
BERING SEA
SEA OF OKHOTSK
Kuril trench
SEA OF JAPAN
Japan trench
EAST CHINA SEA
Ryukyu trench
Marianas trench
PHILIPPINE
SOUTH CHINA SEA
PACIFIC
PACIFIC
OCEAN
CORAL SEA
New Hebrides trench
Tonga trench
TASMAN SEA
Earth's structural regions
earthquake zone
volcano
transform fault
plate boundaries
constructive
destructive
uncertain
© European Map Graphics Ltd.

GLOBAL WARMING
ARCTIC OCEAN
Arctic Circle
BAFFIN BAY
GREENLAND SEA
GULF OF ALASKA
HUDSON BAY
BERING SEA
SEA OF OKHOTSK
SEA OF JAPAN
BLACK SEA
MEDITERRANEAN SEA
PACIFIC OCEAN
Tropic of Cancer
ATLANTIC OCEAN
GULF OF MEXICO
CARIBBEAN SEA
EAST CHINA SEA
SOUTH CHINA SEA
ARABIAN SEA
BAY OF BENGAL
Equator
INDIAN OCEAN
CORAL SEA
Tropic of Capricorn
TASMAN SEA
Carbon dioxide (CO_2) emissions in tonnes per person per year
Over 10 tonnes
5 – 10 tonnes
1 – 5 tonnes
Under 1 tonne
Coastal areas vulnerable to rising sea levels caused by global warming
© European Map Graphics Ltd.

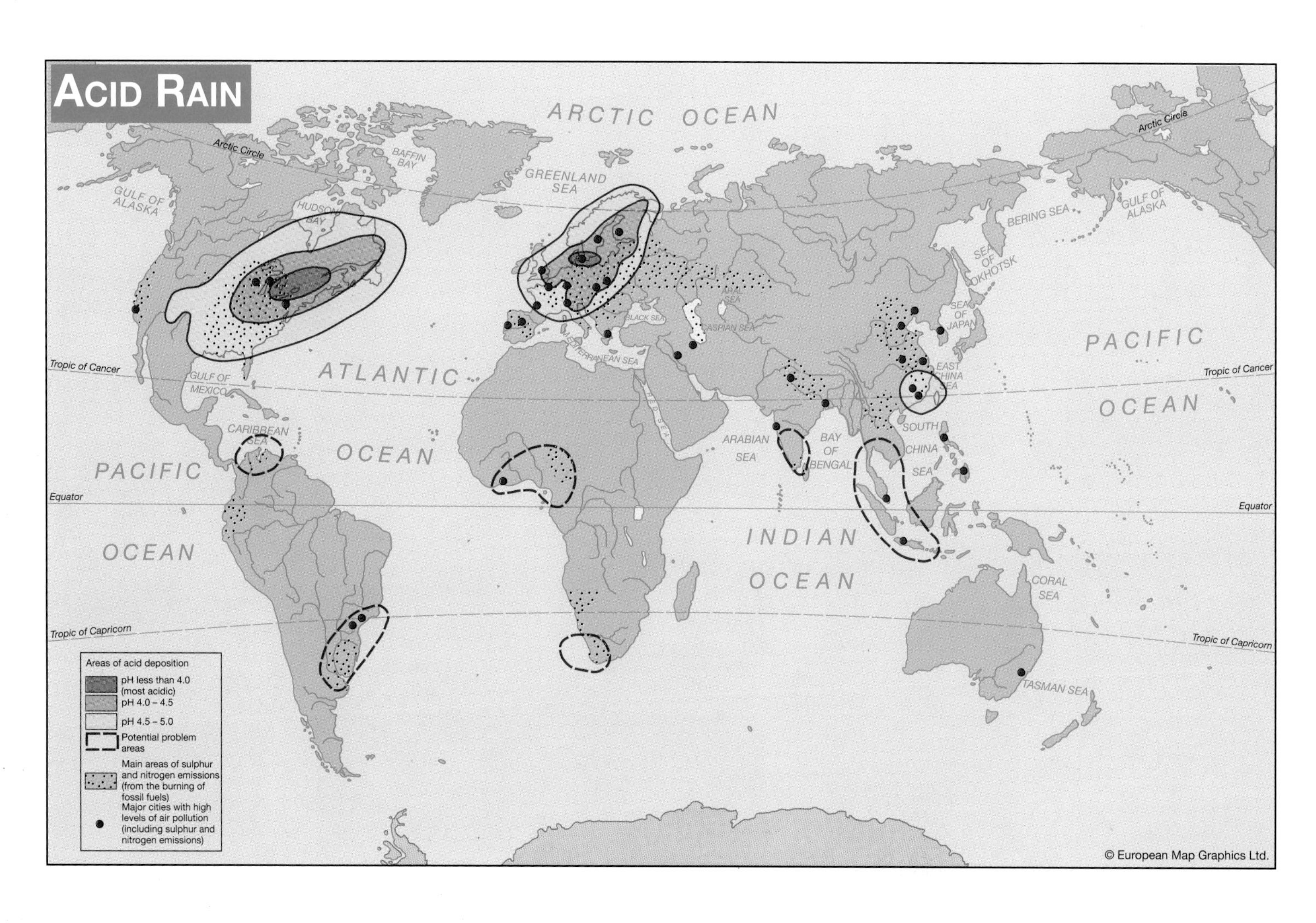
ACID RAIN
ARCTIC OCEAN
Arctic Circle
BAFFIN BAY
GREENLAND SEA
GULF OF ALASKA
HUDSON BAY
BERING SEA
SEA OF OKHOTSK
SEA OF JAPAN
BLACK SEA
CASPIAN SEA
MEDITERRANEAN SEA
PACIFIC OCEAN
Tropic of Cancer
ATLANTIC OCEAN
GULF OF MEXICO
CARIBBEAN SEA
EAST CHINA SEA
SOUTH CHINA SEA
ARABIAN SEA
BAY OF BENGAL
Equator
INDIAN OCEAN
CORAL SEA
Tropic of Capricorn
TASMAN SEA
Areas of acid deposition
pH less than 4.0 (most acidic)
pH 4.0 – 4.5
pH 4.5 – 5.0
Potential problem areas
Main areas of sulphur and nitrogen emissions (from the burning of fossil fuels)
Major cities with high levels of air pollution (including sulphur and nitrogen emissions)
© European Map Graphics Ltd.

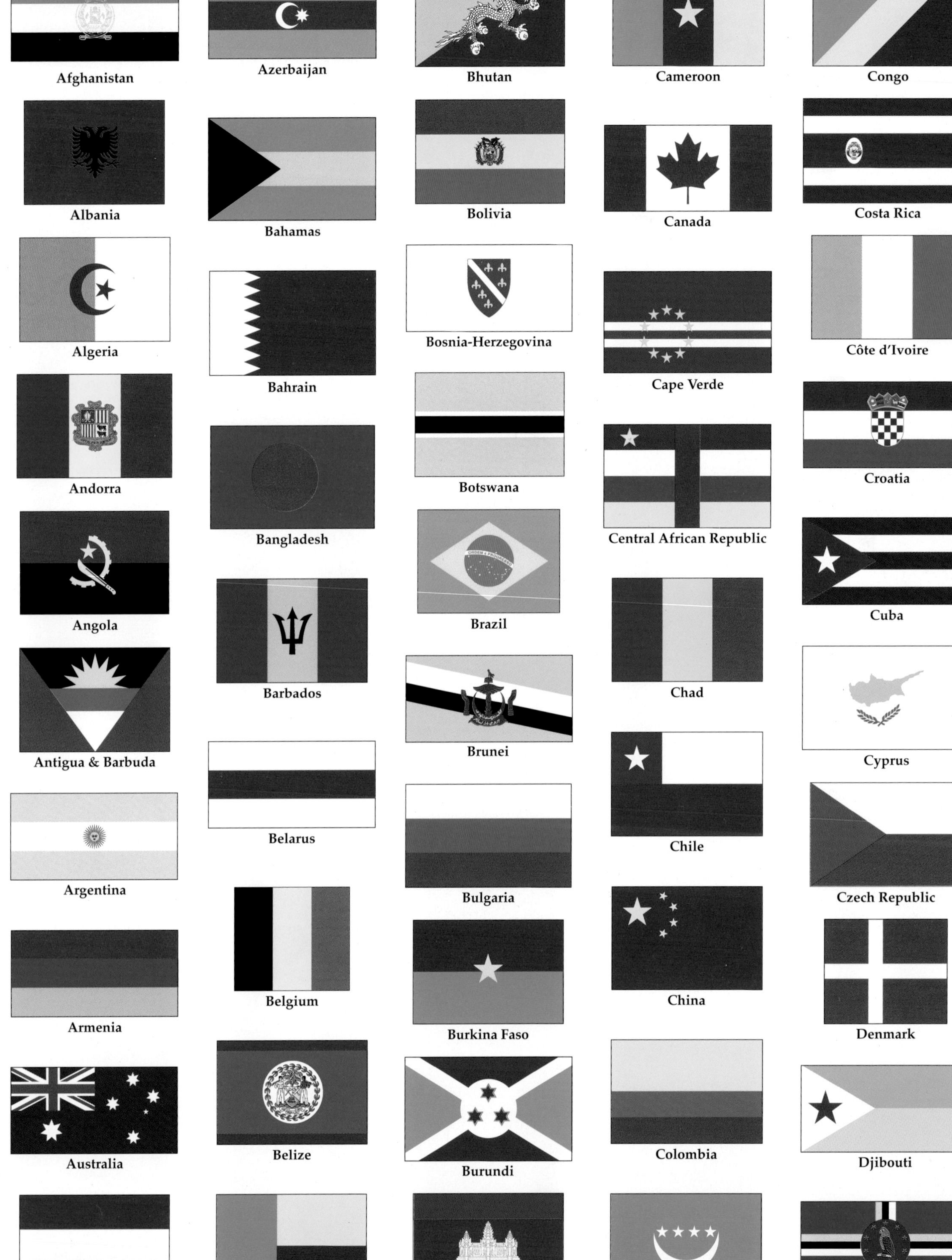

Afghanistan

Albania

Algeria

Andorra

Angola

Antigua & Barbuda

Argentina

Armenia

Australia

Austria

Azerbaijan

Bahamas

Bahrain

Bangladesh

Barbados

Belarus

Belgium

Belize

Benin

Bhutan

Bolivia

Bosnia-Herzegovina

Botswana

Brazil

Brunei

Bulgaria

Burkina Faso

Burundi

Cambodia

Cameroon

Canada

Cape Verde

Central African Republic

Chad

Chile

China

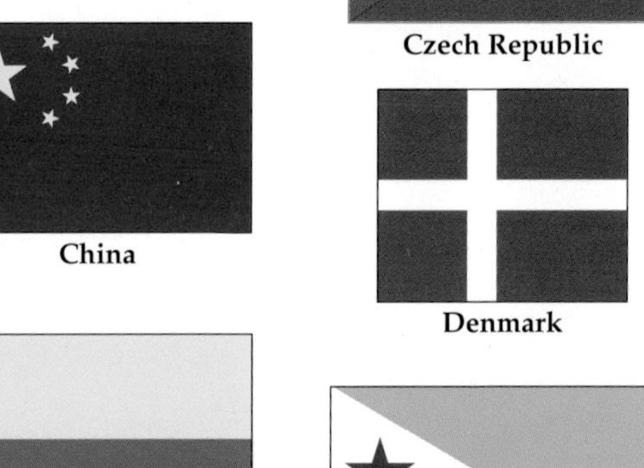

Colombia

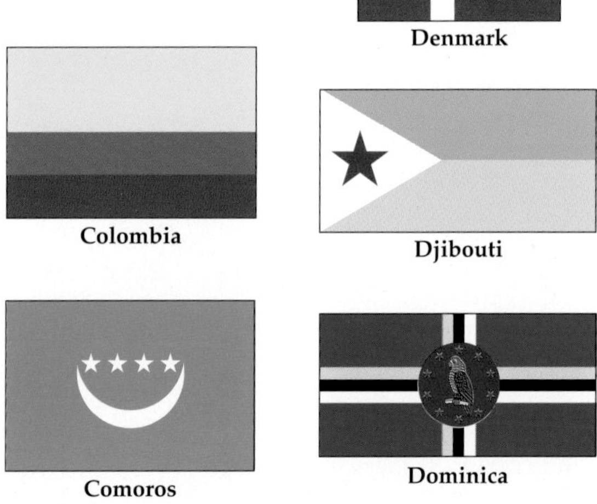

Comoros

Congo

Costa Rica

Côte d'Ivoire

Croatia

Cuba

Cyprus

Czech Republic

Denmark

Djibouti

Dominica

Dominican Republic

Ecuador

Egypt

El Salvador

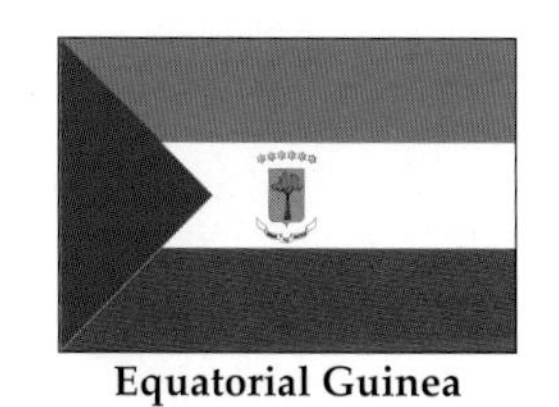
Equatorial Guinea

Eritrea

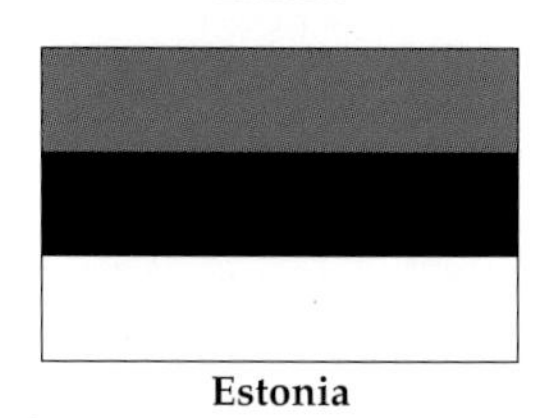
Estonia

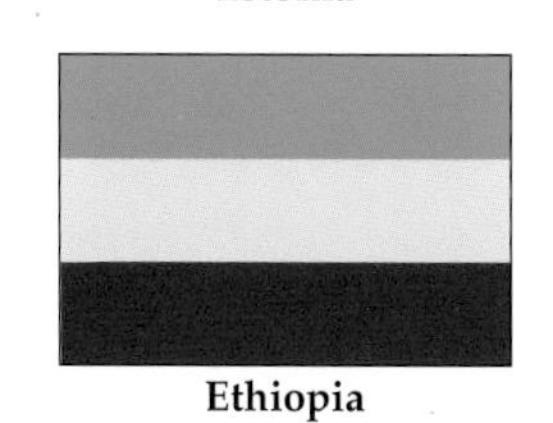
Ethiopia

Fiji

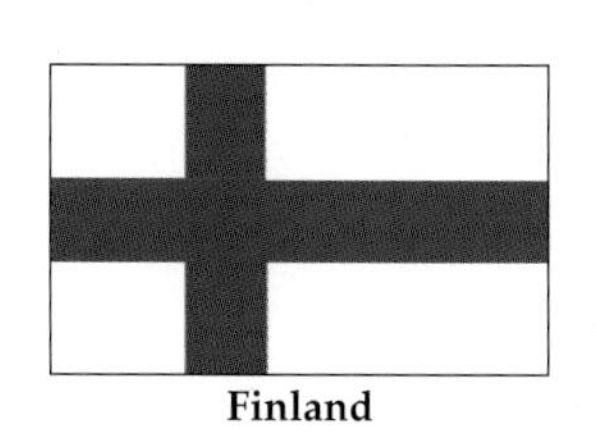
Finland

France

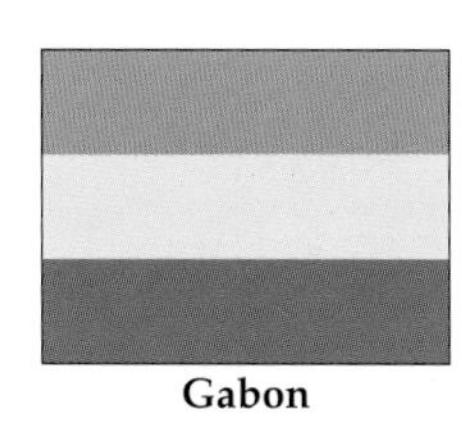
Gabon

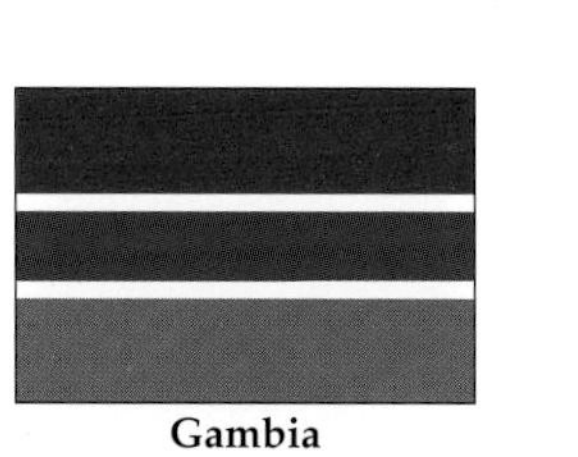
Gambia

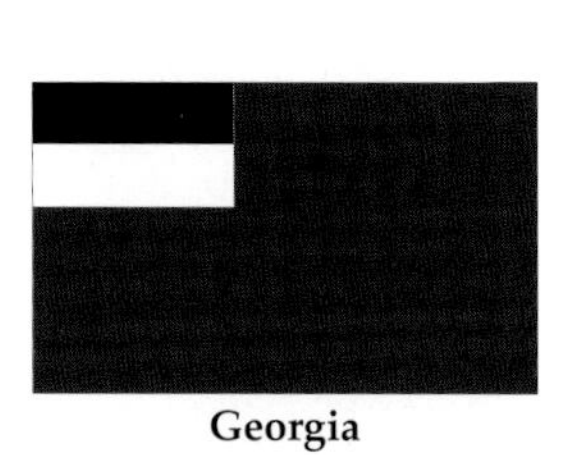
Georgia

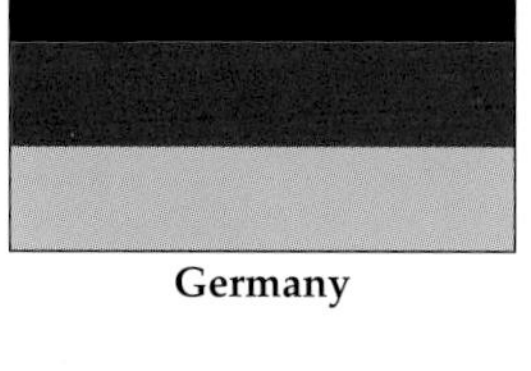
Germany

Ghana

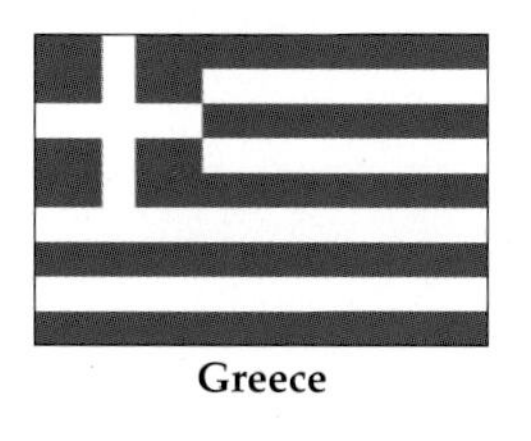
Greece

Grenada

Guatemala

Guinea

Guinea Bissau

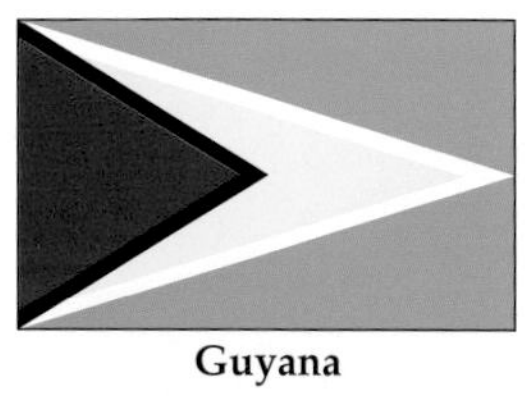
Guyana

Haiti

Honduras

Hungary

Iceland

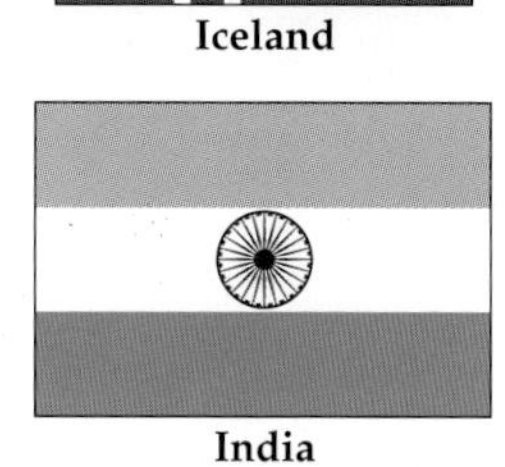
India

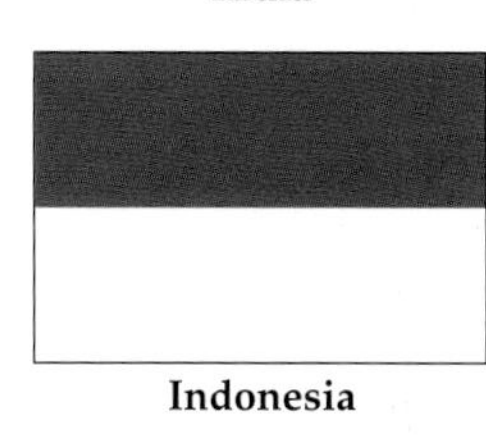
Indonesia

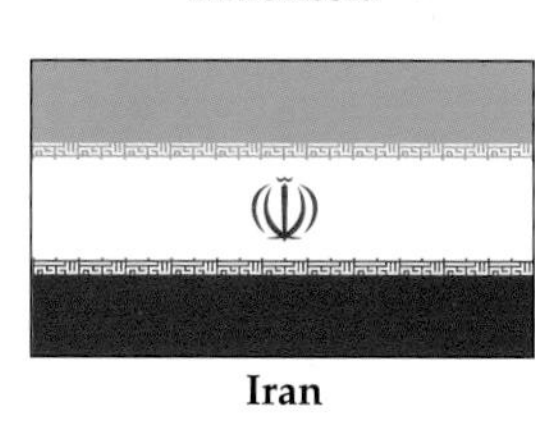
Iran

Iraq

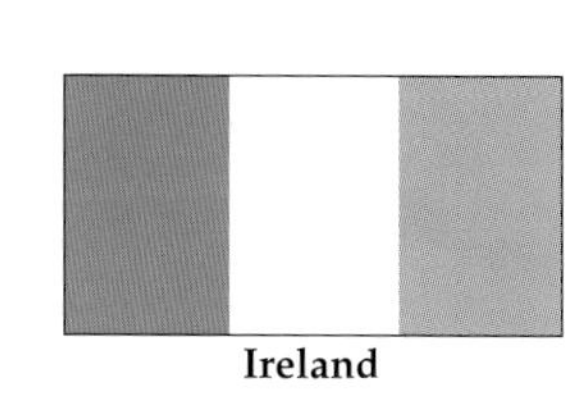
Ireland

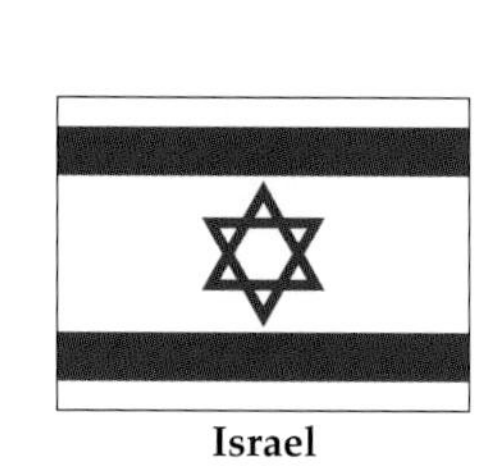
Israel

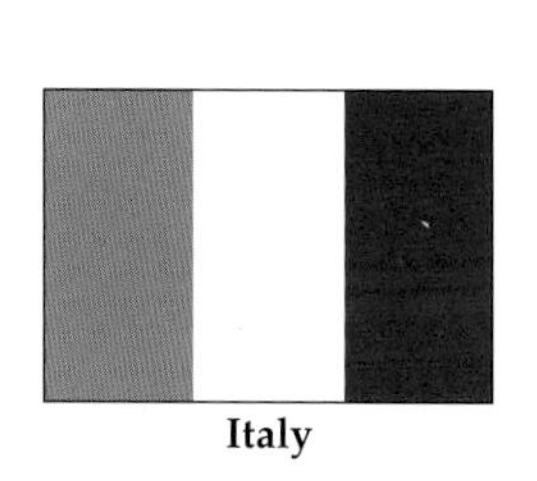
Italy

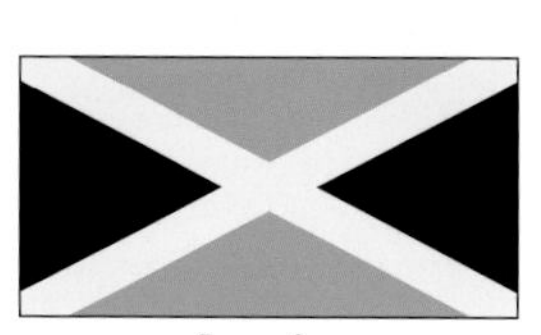

Jamaica

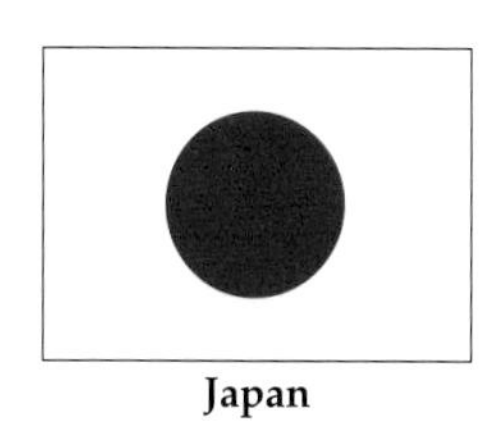
Japan

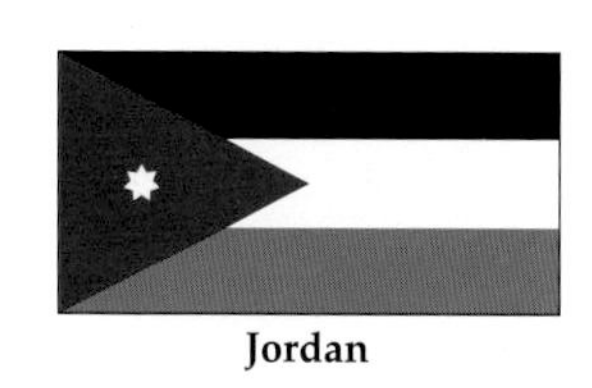
Jordan

Kazakhstan

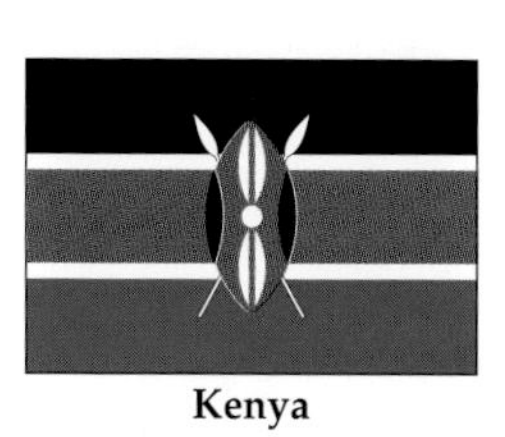
Kenya

Kiribati

Korea, North

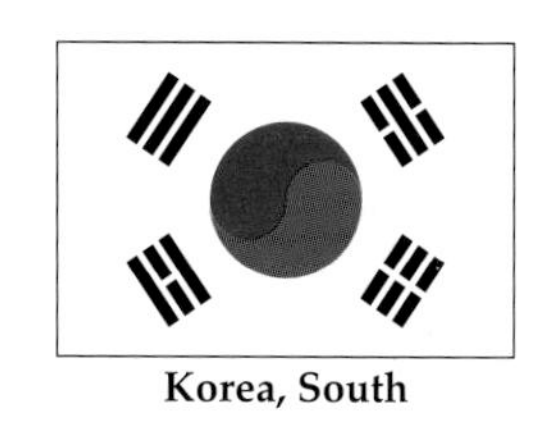
Korea, South

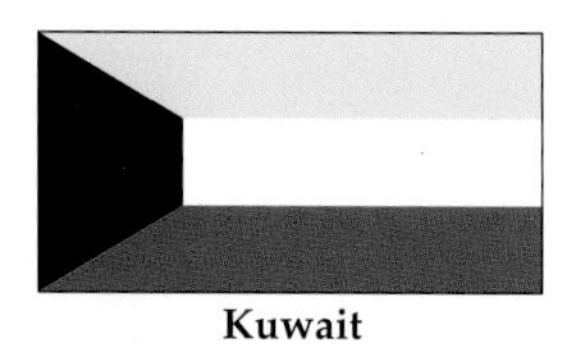
Kuwait

Kyrgyzstan

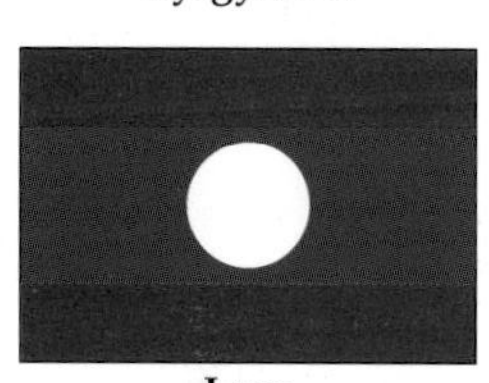
Laos

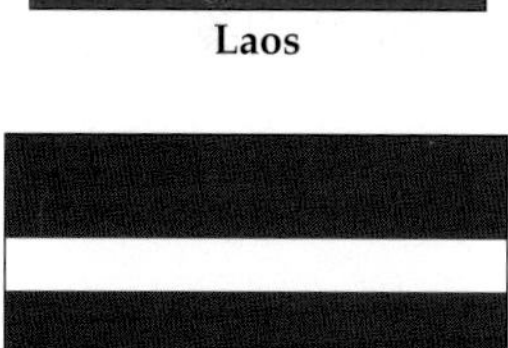
Latvia

Lebanon

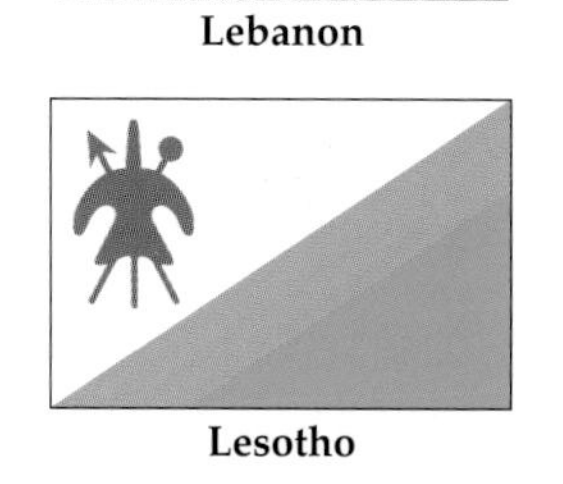
Lesotho

Liberia

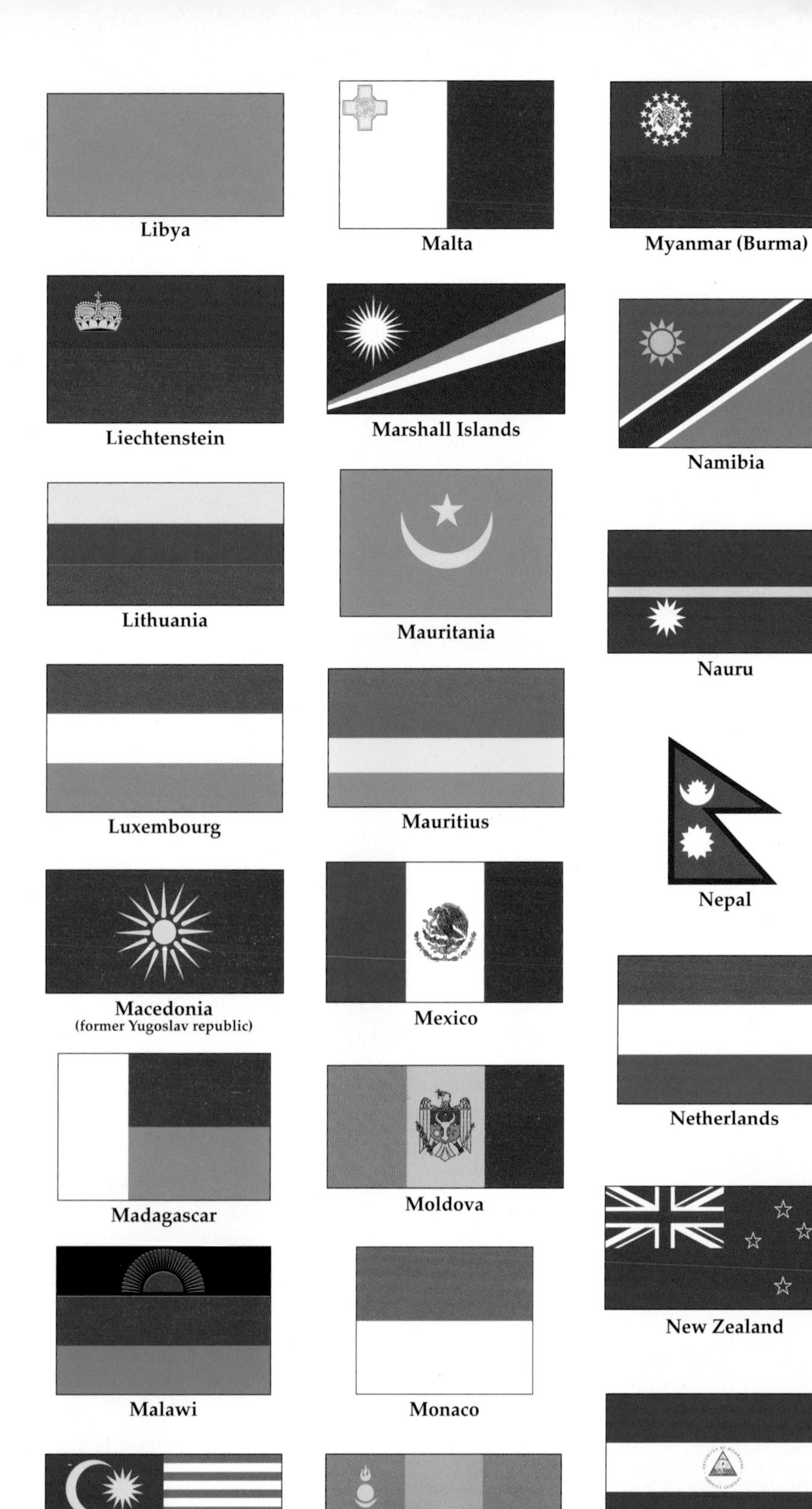

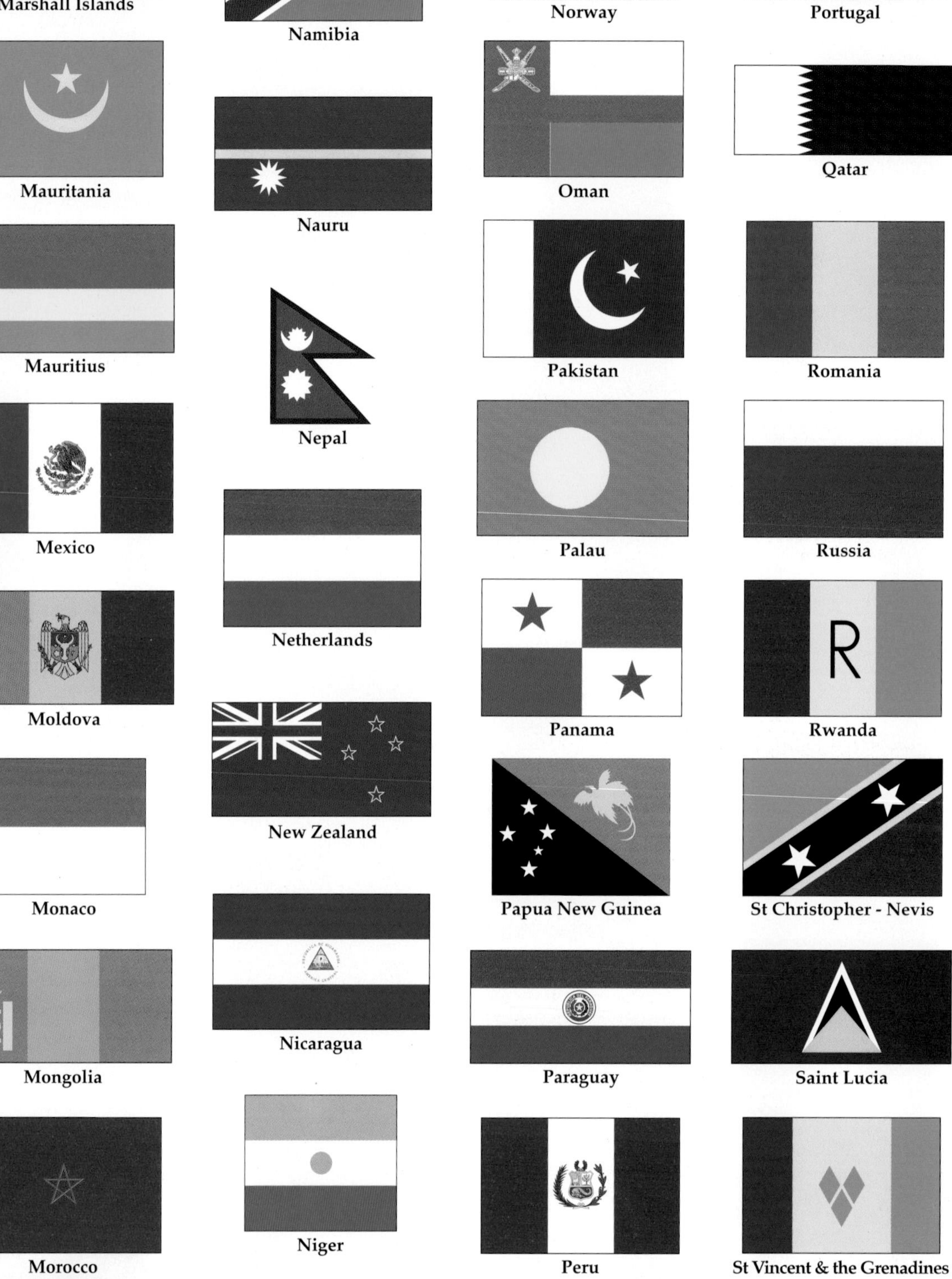

Libya
Malta
Myanmar (Burma)
Northern Marianas
Poland

Liechtenstein
Marshall Islands
Namibia
Norway
Portugal

Lithuania
Mauritania
Nauru
Oman
Qatar

Luxembourg
Mauritius
Nepal
Pakistan
Romania

Macedonia (former Yugoslav republic)
Mexico
Netherlands
Palau
Russia

Madagascar
Moldova
New Zealand
Panama
Rwanda

Malawi
Monaco
Nicaragua
Papua New Guinea
St Christopher - Nevis

Malaysia
Mongolia
Niger
Paraguay
Saint Lucia

Maldives
Morocco
Nigeria
Peru
St Vincent & the Grenadines

Mali
Mozambique
Philippines
San Marino

São Tomé & Príncipe

Saudi Arabia

Senegal

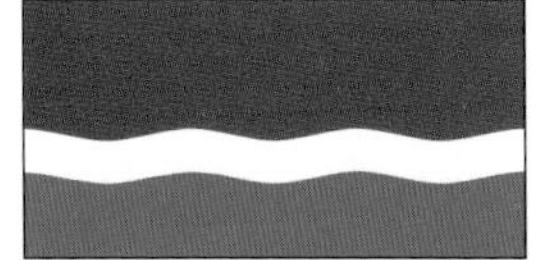
Seychelles

Sierra Leone

Singapore

Slovakia

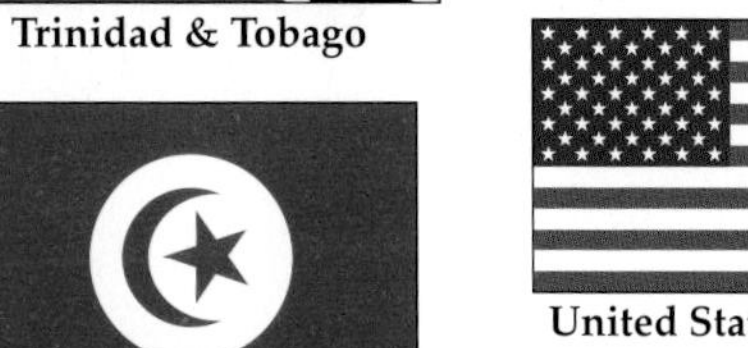
Slovenia

Solomon Islands

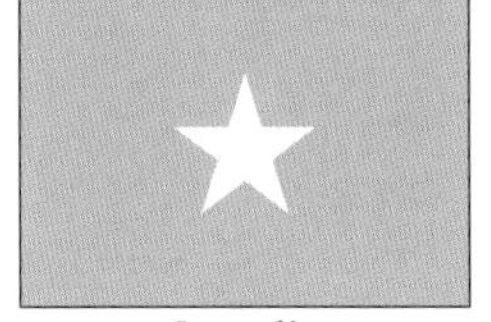
Somalia

South Africa

Spain

Sri Lanka

Sudan

Surinam

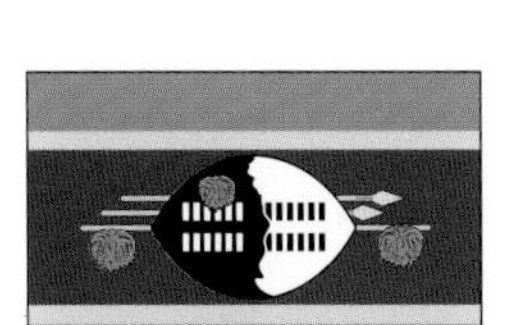
Swaziland

Sweden

Switzerland

Syria

Taiwan

Tajikistan

Tanzania

Thailand

Togo

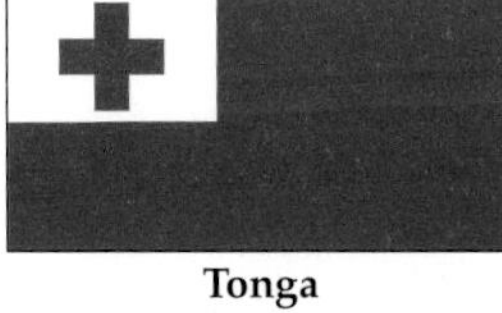
Tonga

Trinidad & Tobago

Tunisia

Turkey

Turkmenistan

Tuvalu

Uganda

Ukraine

United Arab Emirates

United Kingdom

United Nations

United States of America

Uruguay

Uzbekistan

Vanuatu

Vatican City

Venezuela

Vietnam

Western Samoa

Yemen

Yugoslavia

Zaïre

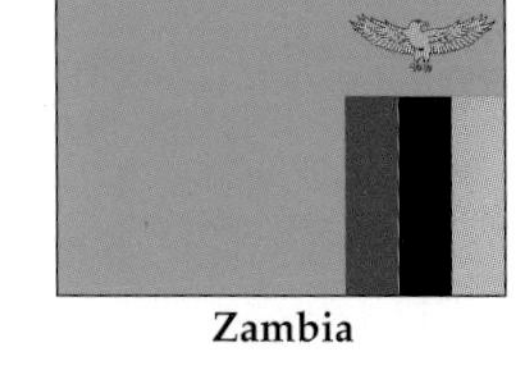
Zambia

Zimbabwe

ICELAND
Reykjavik
NORWEGIAN SEA
BARENTS SEA
0 200 400 600 800 km
Faeroe Is. (Den.)
Shetland Is.
Orkney Is.
Hebrides
NORWAY
SWEDEN
FINLAND
Trondheim
Bergen
Oslo
Stockholm
Helsinki
Tallinn
Tampere
Gulf of Bothnia
Vänern
Vättern
Göteborg
ESTONIA
Riga
LATVIA
LITHUANIA
Vilnius
RUSSIA
Minsk
BELARUS
SCOTLAND
Glasgow
Edinburgh
NORTH SEA
Belfast
N.I.
UNITED KINGDOM
Dublin
IRELAND
Liverpool
Leeds
Manchester
WALES
ENGLAND
Birmingham
London
DENMARK
Copenhagen
Malmö
BALTIC SEA
Kiel
Hamburg
NETHERLANDS
Amsterdam
The Hague
Rotterdam
Hannover
Berlin
POLAND
Poznan
Lódz
Warsaw
Wroclaw
Krakow
ATLANTIC OCEAN
Channel Is.
English Channel
Brussels
BELGIUM
Lille
GERMANY
Essen
Cologne
Bonn
Leipzig
Dresden
Frankfurt
LUXEMBOURG
Luxembourg
Paris
Loire
Rhine
Prague
CZECH REPUBLIC
Stuttgart
Danube
L'vov
Kiev
UKRAINE
Nantes
Bay of Biscay
FRANCE
Munich
Vienna
Bratislava
SLOVAKIA
Zürich
SWITZERLAND
Bern
LIECHTENSTEIN
AUSTRIA
Graz
Budapest
Miskolc
HUNGARY
MOLDOVA
Chişinau
Cluj-Napoca
ROMANIA
Bordeaux
Lyon
Turin
Milan
Venice
SLOVENIA
Ljubljana
Zagreb
CROATIA
Timisoara
Oporto
PORTUGAL
Bilbao
Toulouse
Rhône
Genoa
Bologna
BOSNIA-HERZEGOVINA
Sarajevo
Belgrade
Bucharest
Ebro
ANDORRA
Marseilles
MONACO
Nice
Florence
SAN MARINO
Zaragoza
Madrid
SPAIN
Lisbon
Barcelona
Corsica (France)
ITALY
ADRIATIC SEA
YUGOSLAVIA
Sofia
BULGARIA
Plovdiv
Valencia
Rome
Skopje
Tiranë
MACEDONIA
ALBANIA
Seville
Sardinia (Italy)
Naples
Bari
Thessaloniki
Balearic Is. (Spain)
TYRRHENIAN SEA
Málaga
Gibraltar (U.K.)
IONIAN SEA
GREECE
AEGEAN SEA
MEDITERRANEAN SEA
Palermo
Sicily
Athens
MOROCCO
ALGERIA
TUNISIA
MALTA
Crete
5°West
0°
5°East
10°
15°
20°
25°
30°
35°
40°
45°
50°
55°
60°

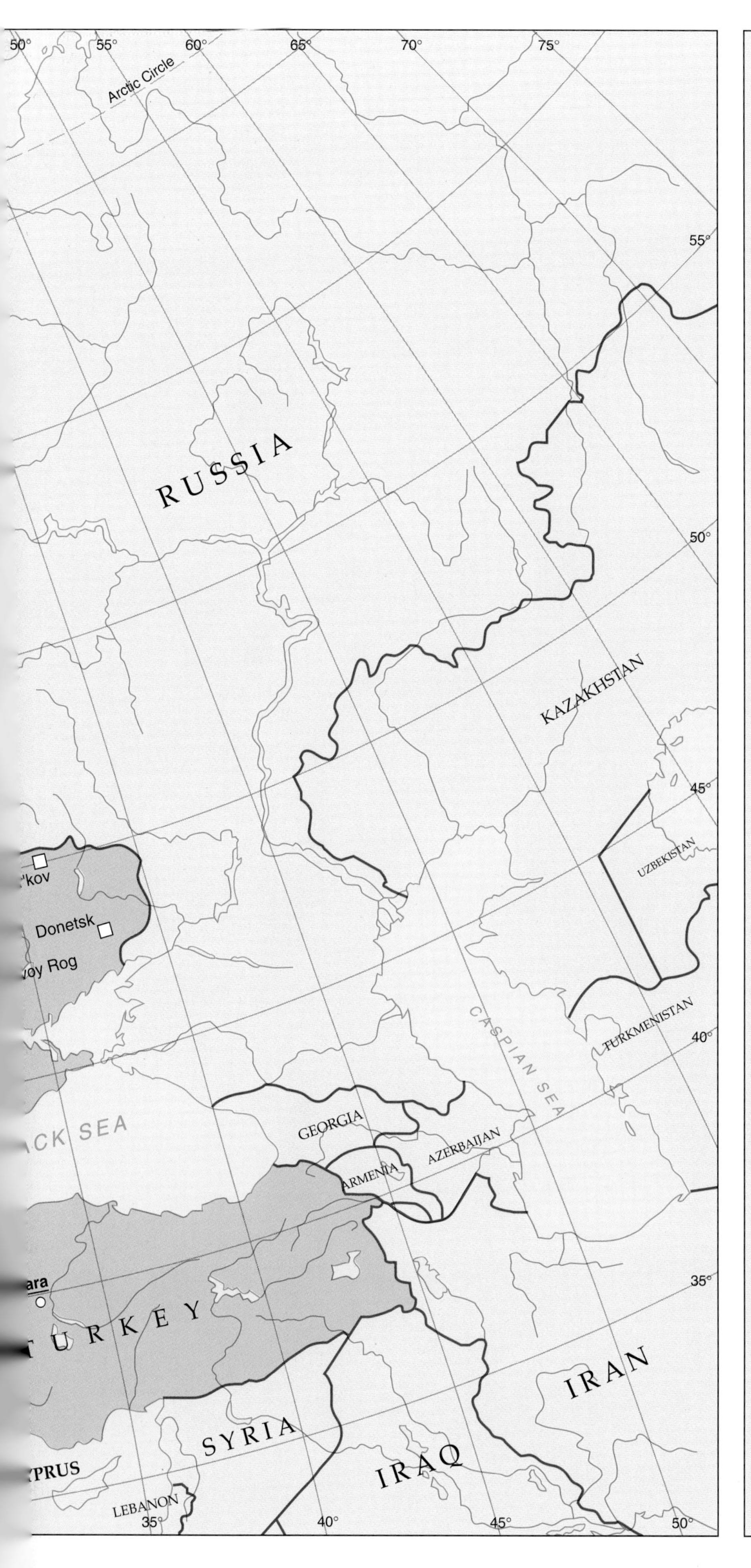
50°
55°
60°
65°
70°
75°
Arctic Circle
55°
50°
45°
40°
35°
RUSSIA
KAZAKHSTAN
UZBEKISTAN
TURKMENISTAN
CASPIAN SEA
'kov
Donetsk
oy Rog
CK SEA
GEORGIA
ARMENIA
AZERBAIJAN
TURKEY
ara
IRAN
SYRIA
IRAQ
PRUS
LEBANON
35°
40°
45°
50°

COUNTRY	PAGE
Albania	16
Andorra	16
Austria	16
Balearic Islands	17
Belarus	17
Belgium	17
Bosnia-Herzegovina	18
Bulgaria	18
Croatia	18
Cyprus	19
Czech Republic	19
Denmark	19
England	30
Estonia	20
Finland	21
France	20
Germany	21
Greece	22
Hungary	22
Iceland	22
Ireland	23
Italy	23
Latvia	24
Liechtenstein	24
Lithuania	24
Luxembourg	24
Macedonia (Former Yugoslav Republic)	25
Malta	25
Moldova	25
Monaco	20
Netherlands	26
Northern Ireland	31
Norway	26
Poland	26
Portugal	27
Romania	27
San Marino	23
Scotland	31
Slovakia	27
Slovenia	28
Spain	28
Sweden	28
Switzerland	29
Turkey	29
Ukraine	29
United Kingdom	30
Vatican City	23
Wales	31
Yugoslavia	29

ALBANIA

Facts and Figures

Capital: Tiranë 251,000 (1990)
Area (sq km): 28,748
Population: 3,300,000 (1991)
Language: Albanian, Greek, English
Religion: Sunni Muslim 70%
Orthodox 20%
Roman Catholic 10%
Currency: Lek = 100 quindars
Annual Income per person: $1,000
Annual Trade per person: $73
Adult Literacy: 85%
Life Expectancy (F): 75
Life Expectancy (M): 70

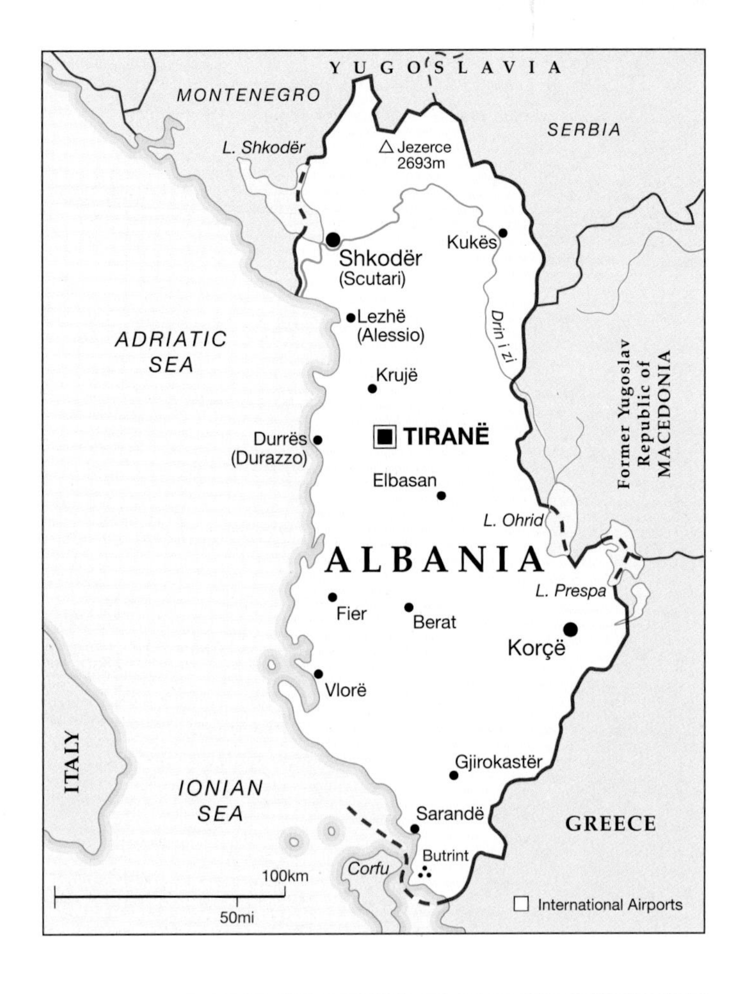

ANDORRA

Facts and Figures

Capital: Andorra la vella 19,000 (1990)
Area (sq km): 453
Population: 59,000 (1993)
Language: Catalan, Spanish, French
Religion: Roman Catholic
Currency: French Franc, Spanish Peseta
Annual Income per person: $14,000
Annual Trade per person: $9,000
Adult Literacy: Data not available
Life Expectancy (F): 81
Life Expectancy (M): 74

AUSTRIA

Facts and Figures

Capital: Vienna 1,539,850 (1991)
Area (sq km): 83,860
Population: 7,800,000 (1991)
Language: German
Religion: Roman Catholic 78%
Protestant 5%
Currency: Schilling = 100 groschen
Annual Income per person: $20,380
Annual Trade per person: $11,653
Adult Literacy: 99%
Life Expectancy (F): 79
Life Expectancy (M): 72

BALEARIC ISLANDS

Cabo de Formentor
Menorca
Mahón
Sierra del Norte
Dragonera I
Mallorca
Palma
Manacor
Bahía de Palma
Cabo de Salinas
Cabrera
Ibiza
Ibiza
Formentera
MEDITERRANEAN SEA
International Airports
100km
50mi

BALEARIC ISLANDS

Facts and Figures

Facts and Figures for the Balearic Islands are incorporated within the details given for Spain

BELARUS

Facts and Figures

Capital: Minsk 1,613,000 (1990)
Area (sq km): 207,600
Population: 10,280,000 (1992)
Language: Belorussian, Russian
Religion: Christian Orthodox, Roman Catholic
Currency: Rouble
Annual Income per person: $3,110
Annual Trade per person: $1,000
Adult Literacy: 95%
Life Expectancy (F): 75
Life Expectancy (M): 64

NORTH SEA
Maas
NETHERLANDS
Zeebrugge
Ostend
Antwerp
Bruges
Ghent
FLANDERS
Mechelen
GERMANY
Schelde
BRUSSELS
BELGIUM
Tournai
Liège
Meuse
Mons
Charleroi
Namur
WALLONIA
Dinant
Ardennes
FRANCE
LUXEMBOURG
Arlon
International Airports
100km
50mi
Moselle

BELGIUM

Facts and Figures

Capital: Brussels 1,331,000 (1991)
Area (sq km): 30,530
Population: 10,020,000 (1992)
Language: Flemish, Dutch, Walloon
Religion: Roman Catholic 72% Protestant
Currency: Belgian Franc
Annual Income per person: $19,300
Annual Trade per person: $23,443
Adult Literacy: 99%
Life Expectancy (F): 79
Life Expectancy (M): 72

BOSNIA-HERZEGOVINA

Facts and Figures

Capital: Sarajevo 526,000 (1991)
Area (sq km): 51,129
Population: 4,366,000 (1991)
Language: Serbo-Croat
Religion: Muslim 40% Orthodox 31%
Roman Catholic 15%
Currency: Dinar
Annual Income per person: $3,000
Annual Trade per person: $900
Adult Literacy: 86%
Life Expectancy (F): 73
Life Expectancy (M): 68

BULGARIA

Facts and Figures

Capital: Sofia 1,141,140 (1990)
Area (sq km): 110,994
Population: 8,470,000 (1992)
Language: Bulgarian, Turkish, Romany
Religion: Eastern Orthodox 80%
Sunni Muslim
Currency: Lev = 100 stotinki
Annual Income per person: $1,840
Annual Trade per person: $2,000
Adult Literacy: 93%
Life Expectancy (F): 76
Life Expectancy (M): 70

CROATIA

Facts and Figures

Capital: Zagreb 726,770 (1991)
Area (sq km): 56,540
Population: 4,790,000 (1992)
Language: Serbo-Croat
Religion: Roman Catholic 77%
Orthodox 11%
Currency: Croatian Dinar
Annual Income per person: $5,600
Annual Trade per person: $1,500
Adult Literacy: 96%
Life Expectancy (F): 74
Life Expectancy (M): 67

CYPRUS

Facts and Figures

Capital: Nicosia 166,500 (1991)
Area (sq km): 9,250
Population: 725,000 (1994)
Language: Greek, Turkish, English
Religion: Greek Orthodox 79%
Muslim 18%
Currency: Cyprus Pound
Annual Income per person: $8,640
Annual Trade per person: $5,032
Adult Literacy: 94%
Life Expectancy (F): 79
Life Expectancy (M): 74

CZECH REPUBLIC

Facts and Figures

Capital: Prague 1,120,000 (1990)
Area (sq km): 79,000
Population: 10,330,000 (1993)
Language: Czech
Religion: Protestant
Currency: Koruna = 100 halura
Annual Income per person: N/A
Annual Trade per person: N/A
Adult Literacy: 99%
Life Expectancy (F): 76
Life Expectancy (M): 69

DENMARK

Facts and Figures

Capital: Copenhagen 1,342,680 (1993)
Area (sq km): 43,075
Population: 5,180,000 (1993)
Language: Danish
Religion: Lutheran 90%
Currency: Krone = 100 øre
Annual Income per person: $23,660
Annual Trade per person: $14,046
Adult Literacy: 99%
Life Expectancy (F): 79
Life Expectancy (M): 73

ESTONIA

Facts and Figures

Capital: Tallinn 502,400 (1991)
Area (sq km): 45,100
Population: 1,600,000 (1992)
Language: Estonian, Russian
Religion: Lutheran, Orthodox
Currency: Kroon = 100 sents
Annual Income per person: $3,830
Annual Trade per person: $850
Adult Literacy: N/A
Life Expectancy (F): 75
Life Expectancy (M): 66

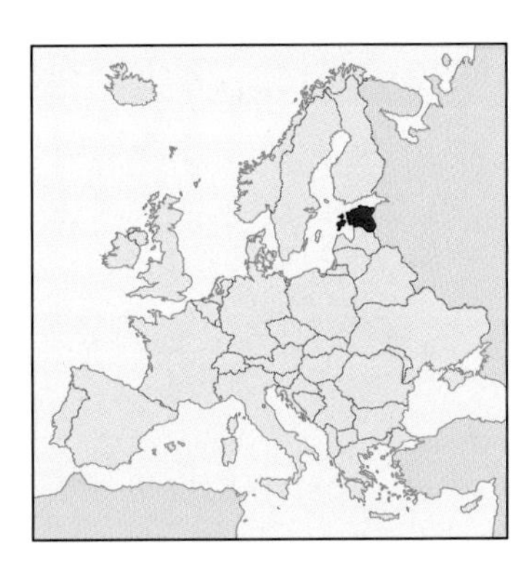

FRANCE

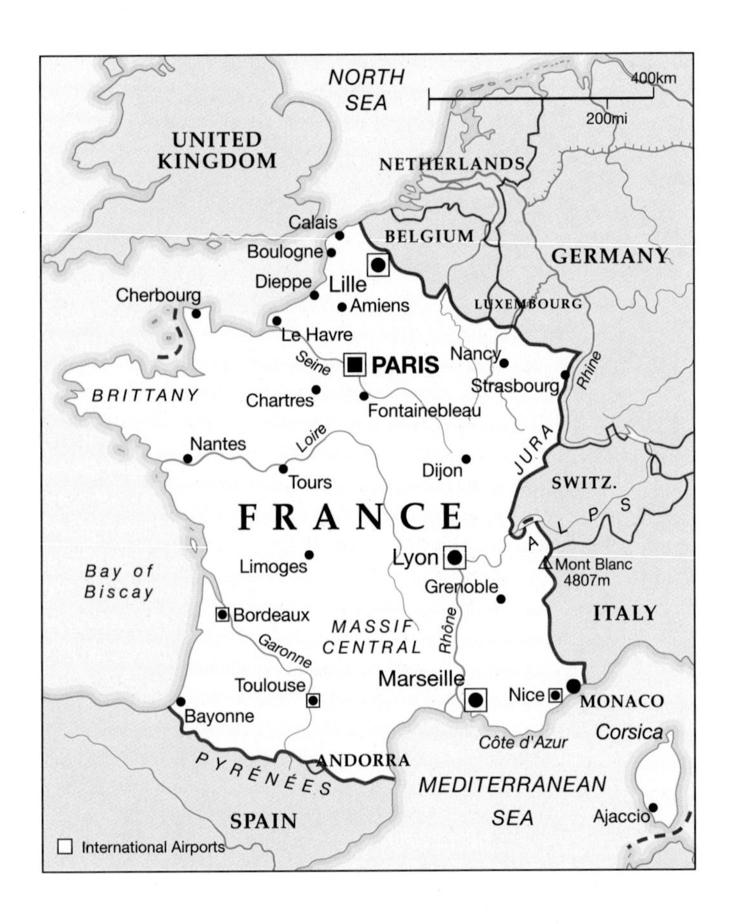

Capital: Paris 9,318,800 (1990)
Area (sq km): 543,965
Population: 57,800,000 (1994)
Language: French
Religion: Roman Catholic 73%
Other Christian 4% Muslim 3%
Currency: Franc = 100 centimes
Annual Income per person: $20,600
Annual Trade per person: $8,195
Adult Literacy: 99%
Life Expectancy (F): 81
Life Expectancy (M): 73

MONACO

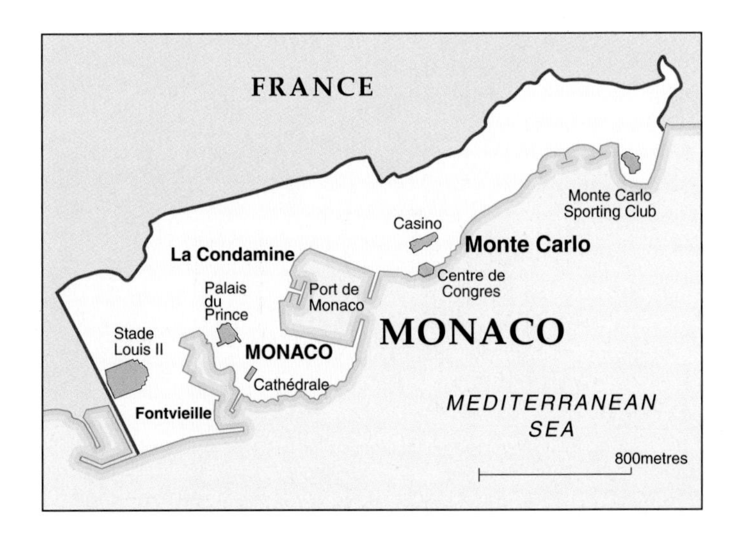

Facts and Figures

Capital: Monaco
Area (sq km): 1.5
Population: 30,000
Language: French, Monégasque
Religion: Roman Catholic 95%
Currency: French currency
Annual Income per person: $16,000
Annual Trade per person: N/A
Adult Literacy: N/A
Life Expectancy (F): 80
Life Expectancy (M): 72

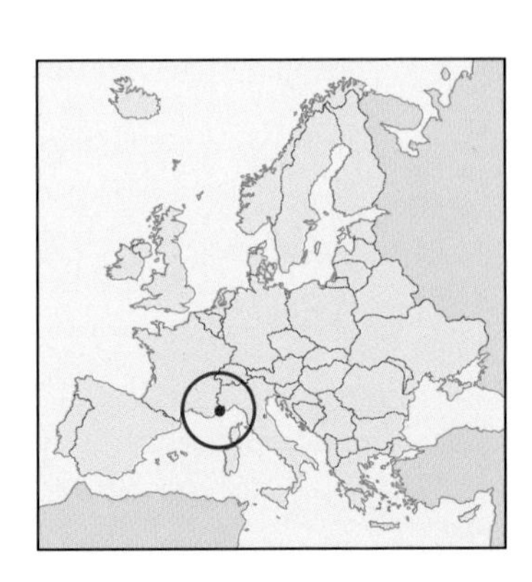

FINLAND

BARENTS SEA
NORWAY
Haltiatunturi 1328m
Lake Inari
Muonio
Sodankyla
Kemi
LAPPLAND
SWEDEN
Arctic Circle
Rovaniemi
Tornio
400km
200mi
International Airports
Oulu
Lake Oulu
RUSSIA
FINLAND
KARELIA
Vaasa
Kuopio
Joensuu
Jyväskylä
Gulf of Bothnia
Lake Saimaa
Tampere
Pori
Lahti
Hämeenlinna
Lake Ladoga
Turku
Vantaa
Espoo
HELSINKI
Åland Is
Gulf of Finland
St Petersburg
ESTONIA

Facts and Figures

Capital: Helsinki 501,500 (1992)
Area (sq km): 338,100
Population: 5,050,000 (1992)
Language: Finnish, Swedish
Religion: Lutheran 87%
Currency: Markka = 100 penniä
Annual Income per person: $21,009
Annual Trade per person: $8,852
Adult Literacy: 99%
Life Expectancy (F): 80
Life Expectancy (M): 72

GERMANY

Facts and Figures

Capital: Berlin 3,437,900 (1990)
Area (sq km): 356,700
Population: 80,280,000 (1992)
Language: German
Religion: Protestant 45%
Roman Catholic 35%
Currency: Deutschmark = 100 pfennig
Annual Income per person: $23,650
Annual Trade per person: $12,860
Adult Literacy: 99%
Life Expectancy (F): 78
Life Expectancy (M): 72

German Regions

GREECE

Facts and Figures

Capital: Athens 3,096,800 (1991)
Area (sq km): 131,960
Population: 10,260,000 (1991)
Language: Greek
Religion: Greek Orthodox 98%
Currency: Drachma = 100 lepta
Annual Income per person: $6,374
Annual Trade per person: $3,005
Adult Literacy: 93%
Life Expectancy (F): 79
Life Expectancy (M): 74

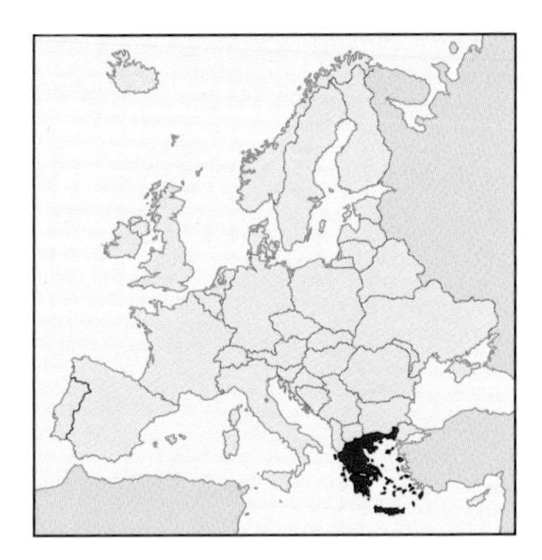

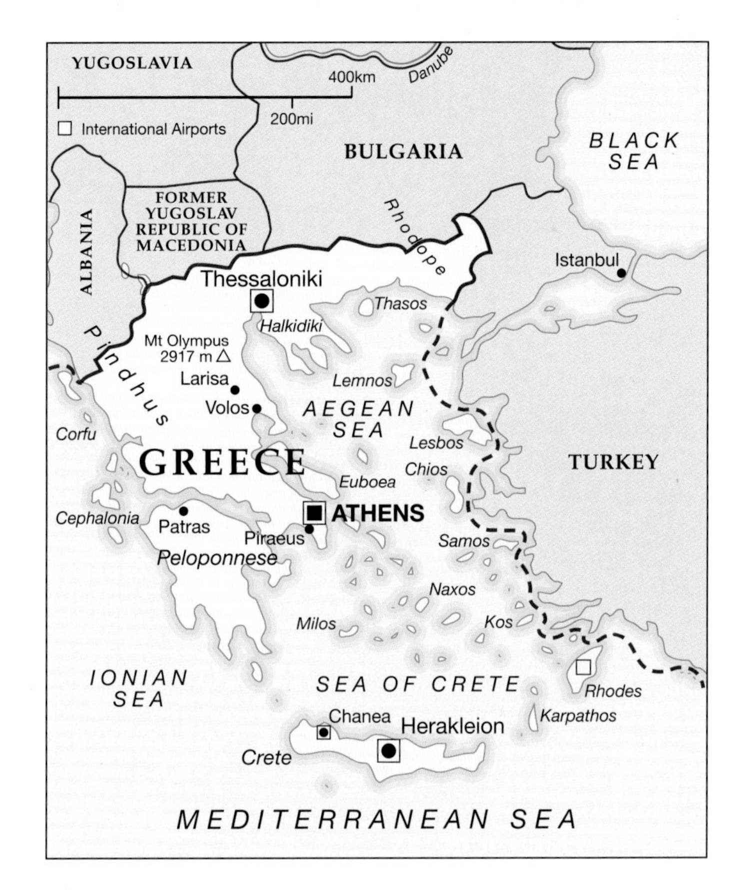

HUNGARY

Facts and Figures

Capital: Budapest 2,016,000 (1992)
Area (sq km): 93,030
Population: 10,310,000 (1993)
Language: Hungarian, Slovak, German
Religion: Roman Catholic 68%
Protestant 25%
Currency: Forint = 100 filler
Annual Income per person: $2,690
Annual Trade per person: $2,109
Adult Literacy: 99%
Life Expectancy (F): 75
Life Expectancy (M): 68

ICELAND

Facts and Figures

Capital: Reykjavik 100,850 (1992)
Area (sq km): 103,000
Population: 262,190 (1992)
Language: Icelandic
Religion: Lutheran 92%
Currency: Króna = 100 aurar
Annual Income per person: $24,570
Annual Trade per person: $12,592
Adult Literacy: 99%
Life Expectancy (F): 81
Life Expectancy (M): 75

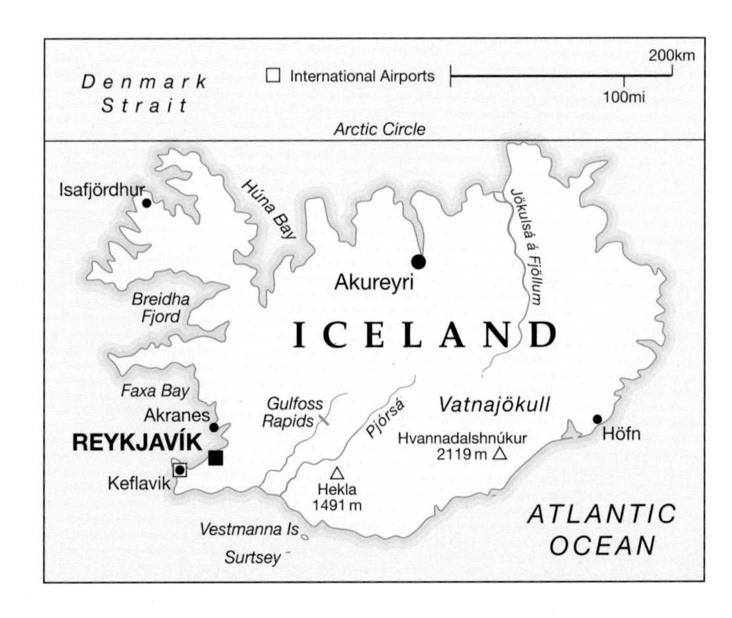

IRELAND

Facts and Figures

Capital: Dublin 915,516 (1991)
Area (sq km): 70,280
Population: 3,550,000 (1992)
Language: Irish, English
Religion: Roman Catholic 90%
Protestant 3%
Currency: Punt = 100 pence
Annual Income per person: $10,780
Annual Trade per person: $14,305
Adult Literacy: 99%
Life Expectancy (F): 78
Life Expectancy (M): 73

ITALY

Facts and Figures

Area (sq km): 0.44
Population: 1,000
Language: Latin, Italian
Religion: Roman Catholic
Currency: Vatican Lira
Annual Income per person: N/A
Annual Trade per person: N/A
Adult Literacy: N/A
Life Expectancy (F): N/A
Life Expectancy (M): N/A

Facts and Figures

Capital: San Marino
Area (sq km): 61
Population: 24,000
Language: Italian
Religion: Roman Catholic
Currency: Lira
Annual Income per person: $17,000
Annual Trade per person: N/A
Adult Literacy: 96%
Life Expectancy (F): 79
Life Expectancy (M): 74

Facts and Figures

Capital: Rome 2,723,330 (1992)
Area (sq km): 301,300
Population: 56,960,000 (1992)
Language: Italian
Religion: Roman Catholic 84%
Currency: Lira
Annual Income per person: $18,580
Annual Trade per person: $6,192
Adult Literacy: 97%
Life Expectancy (F): 80
Life Expectancy (M): 73

LATVIA

Facts and Figures

Capital: Riga 910,200 (1991)
Area (sq km): 63,700
Population: 2,610,000 (1993)
Language: Latvian, Russian
Religion: Lutheran
Currency: Lat = 100 santims
Annual Income per person: $3,410
Annual Trade per person: $3,400
Adult Literacy: Data not available
Life Expectancy (F): 75
Life Expectancy (M): 64

LITHUANIA

Facts and Figures

Capital: Vilnius 592,500 (1990)
Area (sq km): 65,200
Population: 3,740,000 (1994)
Language: Lithuanian, Russian
Religion: Roman Catholic 90%, Lutheran
Currency: Litas = 100 centas
Annual Income per person: $2,710
Annual Trade per person: N/A
Adult Literacy: N/A
Life Expectancy (F): 76
Life Expectancy (M): 66

LIECHTENSTEIN

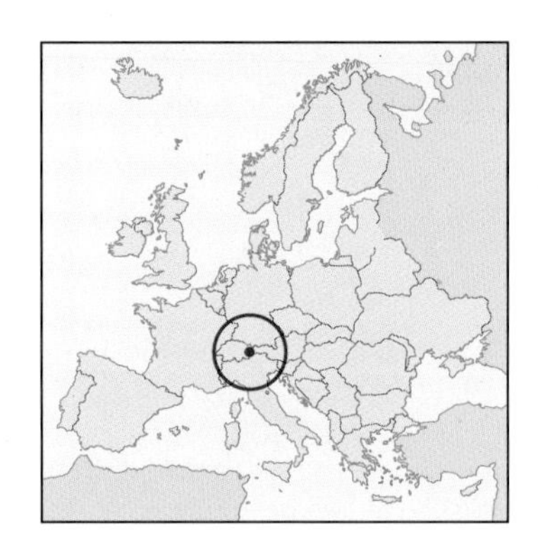

Facts and Figures

Capital: Vaduz
Area (sq km): 157
Population: 30,000
Language: German
Religion: Roman Catholic
Currency: Swiss currency
Annual Income per person: $33,000
Annual Trade per person: N/A
Adult Literacy: 99%
Life Expectancy (F): 81
Life Expectancy (M): 73

LUXEMBOURG

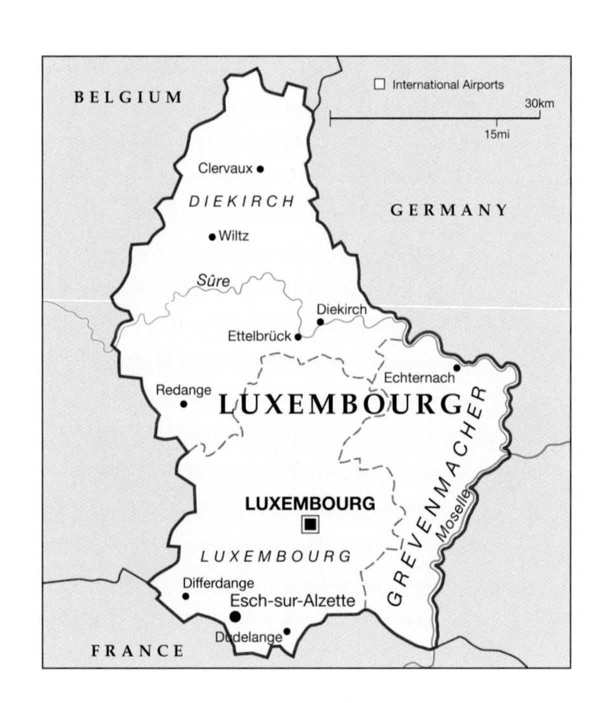

Facts and Figures

Capital: Luxembourg 75,377 (1991)
Area (sq km): 2,590
Population: 395,200 (1993)
Language: Letzeburgish, French, German
Religion: Roman Catholic 95%
Currency: Luxembourg Franc = 100 centimes
Annual Income per person: $31,080
Annual Trade per person: $35,000
Adult Literacy: 100%
Life Expectancy (F): 79
Life Expectancy (M): 72

MACEDONIA

Facts and Figures

Capital: Skopje 448,230 (1991)
Area (sq km): 25,700
Population: 2,060,000 (1992)
Language: Macedonian, Albanian
Religion: Orthodox 67% Muslim 30%
Currency: Dinar = 100 deni
Annual Income per person: $3,100
Annual Trade per person: $800
Adult Literacy: 89%
Life Expectancy (F): 72
Life Expectancy (M): 68

MALTA

Facts and Figures

Capital: Valletta 101,750 (1990)
Area (sq km): 316
Population: 364,600 (1993)
Language: Maltese, English
Religion: Roman Catholic 98%
Currency: Maltese Lira = 100 cents
Annual Income per person: $7,341
Annual Trade per person: $9,311
Adult Literacy: 85%
Life Expectancy (F): 76
Life Expectancy (M): 72

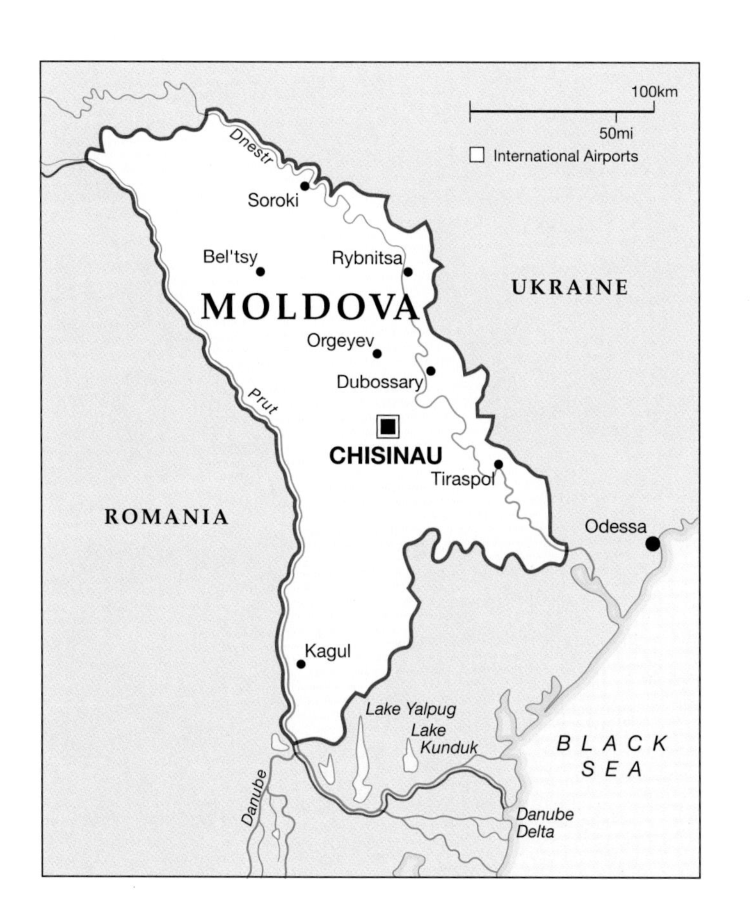

MOLDOVA

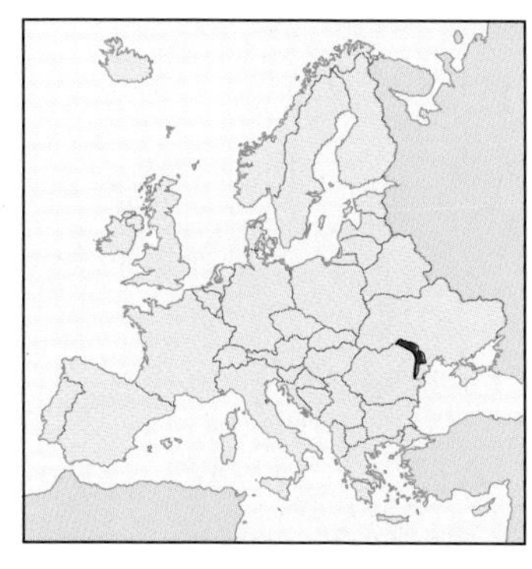

Facts and Figures

Capital: Chisinau 676,000 (1990)
Area (sq km): 33,700
Population: 4,400,000 (1992)
Language: Moldavian, Russian
Religion: Russian Orthodox, Evangelical
Currency: Leu
Annual Income per person: $2,170
Annual Trade per person: N/A
Adult Literacy: N/A
Life Expectancy (F): 72
Life Expectancy (M): 65

NETHERLANDS

Facts and Figures

Capital: Amsterdam 719,860 (1993)
Area (sq km): 41,530
Population: 15,240,000 (1993)
Language: Dutch
Religion: Roman Catholic 33%
Protestant 23%
Currency: Guilder = 100 cents
Annual Income per person: $21,030
Annual Trade per person: $18,137
Adult Literacy: 99%
Life Expectancy (F): 81
Life Expectancy (M): 74

NORWAY

Facts and Figures

Capital: Oslo 473,350 (1992)
Area (sq km): 323,880
Population: 4,300,000 (1992)
Language: Norwegian, Lappish, Finnish
Religion: Lutheran 88% Roman Catholic
Currency: Norwegian Krone = 100 øre
Annual Income per person: $24,160
Annual Trade per person: $14,125
Adult Literacy: 99%
Life Expectancy (F): 81
Life Expectancy (M): 74

POLAND

Facts and Figures

Capital: Warsaw 1,655,000 (1989)
Area (sq km): 312,680
Population: 38,310,000 (1993)
Language: Polish
Religion: Roman Catholic 94%
Currency: Zloty = 100 groszy
Annual Income per person: $1,830
Annual Trade per person: $751
Adult Literacy: 99%
Life Expectancy (F): 76
Life Expectancy (M): 67

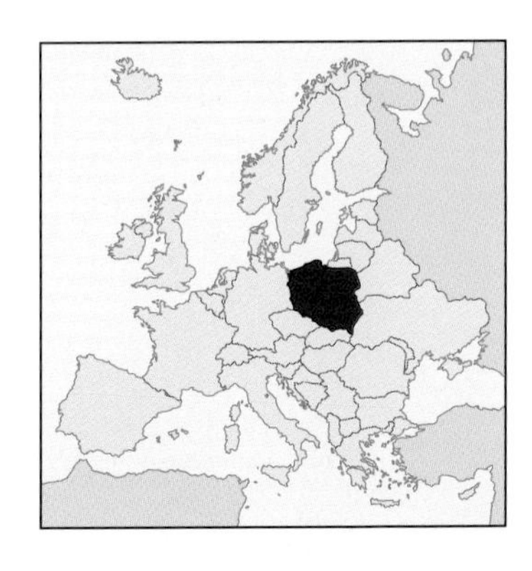

Minho
International Airports
200km
100mi
Costa Verde
Braga
Bragança
Oporto
Douro
ATLANTIC
OCEAN
Salamanca
Aveiro
Serra da Estrêla
Coimbra
Covilhã
Castelo Branco
Tagus
Santarém
Portalegre
SPAIN
PORTUGAL
LISBON
Setúbal
Évora
Guadiana
Beja
Sines
Seville
Lagos
Faro
Cabo de São Vicente
Algarve
Gulf of Cádiz
Doñana
Las Marismas

PORTUGAL

Facts and Figures

Capital: Lisbon 830,500 (1987)
Area (sq km): 91,900
Population: 9,860,000 (1991)
Language: Portuguese
Religion: Roman Catholic
Currency: Escudo = 100 centavos
Annual Income per person: $5,620
Annual Trade per person: $4,836
Adult Literacy: 98%
Life Expectancy (F): 78
Life Expectancy (M): 71

ROMANIA

Facts and Figures

Capital: Bucharest 2,351,000 (1992)
Area (sq km): 237,500
Population: 22,760,000 (1992)
Language: Romanian, Hungarian, German
Religion: Romanian Orthodox 87% RC 5%
Currency: Leu = 100 bani
Annual Income per person: $1,340
Annual Trade per person: $680
Adult Literacy: 96%
Life Expectancy (F): 74
Life Expectancy (M): 69

CZECH REPUBLIC
POLAND
Carpathian Mountains
Žilina
Tatra
Gerlachovsky 2655m
Prešov
Morava
SLOVAK REPUBLIC
Košice
Banská Bystrica
Vah
UKRAINE
Nitra
AUSTRIA
BRATISLAVA
Tisza
Danube
HUNGARY
International Airports
BUDAPEST
200km
100mi

SLOVAK REPUBLIC

Facts and Figures

Capital: Bratislava 440,421 (1990)
Area (sq km): 49,035
Population: 5,334,000 (1992)
Language: Slovak
Religion: Roman Catholic 22%
Currency: Slovak Koruna
Annual Income per person: N/A
Annual Trade per person: N/A
Adult Literacy: 99%
Life Expectancy (F): 76
Life Expectancy (M): 69

SLOVENIA

Facts and Figures

Capital: Ljubljana 268,000 (1991)
Area (sq km): 20,251
Population: 2,000,000 (1992)
Language: Slovene, Serbo-Croat
Religion: Roman Catholic 94%
Currency: Slovene Tolar
Annual Income per person: $10,000
Annual Trade per person: $5,000
Adult Literacy: 99%
Life Expectancy (F): 75
Life Expectancy (M): 67

SPAIN

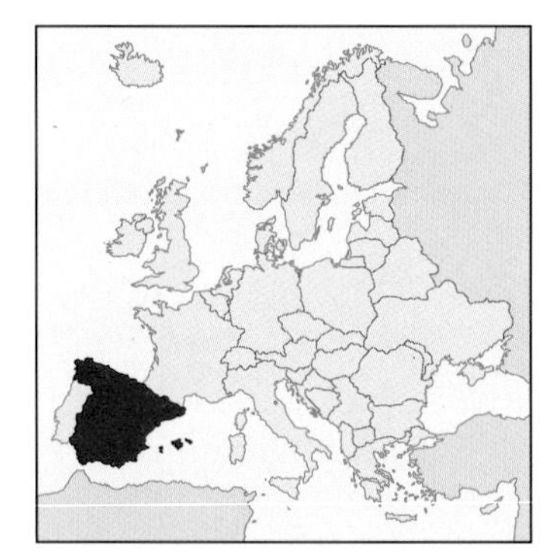

Facts and Figures

Capital: Madrid 2,909,800 (1991)
Area (sq km): 504,750
Population: 39,080,000 (1992)
Language: Castilian Spanish, Basque, Catalan, Galician
Religion: Roman Catholic 97%
Currency: Peseta
Annual Income per person: $14,290
Annual Trade per person: $3,934
Adult Literacy: 96%
Life Expectancy (F): 80
Life Expectancy (M): 74

SWEDEN

Facts and Figures

Capital: Stockholm 684,580 (1992)
Area (sq km): 449,960
Population: 8,700,000 (1992)
Language: Swedish, Finnish
Religion: Lutheran 89%
Currency: Krona = 100 öre
Annual Income per person: $25,490
Annual Trade per person: $12,158
Adult Literacy: 99%
Life Expectancy (F): 81
Life Expectancy (M): 75

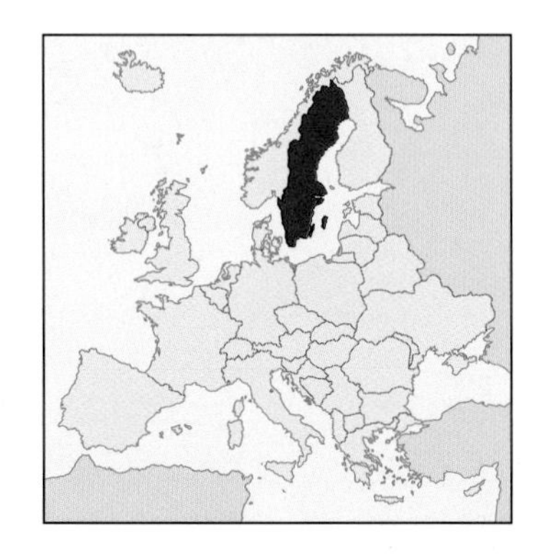

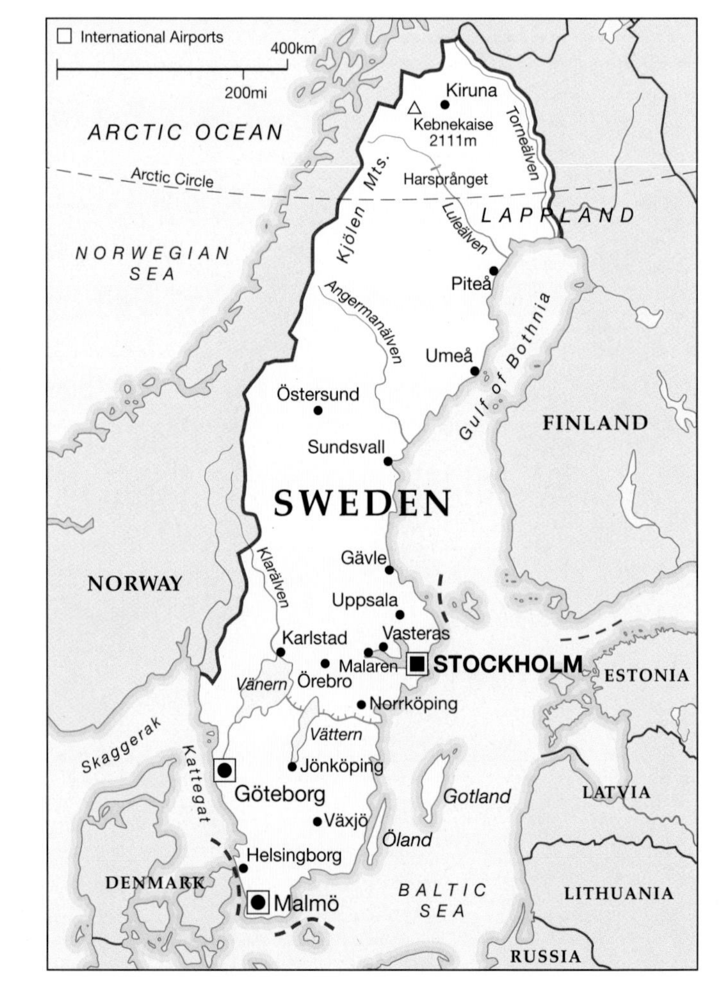

SWITZERLAND

SWITZERLAND
GERMANY
Bodensee (Lake Constance)
FRANCE
Basel
Winterthur
Säntis
St Gallen
Jura Mts.
Aare
Zürich
AUSTRIA
LIECH.
Neuchâtel
Luzern
Vierwald-Stätter See
Lac de Neuchâtel
BERN
Rhine
Thuner See
Andermatt
Davos
Inn
Lausanne
Interlaken
St Moritz
Lac Léman (Lake Geneva)
Montreux
Rhône
S
P
Matterhorn 4477m
L
A
Lugano
Adda
ITALY
Geneva
Lago Maggiore
Lago di Como
Mont Blanc 4807m
Dufourspitze 4634m
150km
100mi
International Airports

Facts and Figures

Capital: Bern 298,700 (1990)
Area (sq km): 41,130
Population: 6,900,000 (1993)
Language: German, French, Italian, Romansch
Religion: Protestant 47%
Roman Catholic 46%
Currency: Swiss Franc = 100 centimes
Annual Income per person: $32,250
Annual Trade per person: $19,088
Adult Literacy: 99%
Life Expectancy (F): 81
Life Expectancy (M): 75

TURKEY

Facts and Figures

Capital: Ankara 3,022,000 (1990)
Area (sq km): 779,450
Population: 59,870,000 (1993)
Language: Turkish, Kurdish
Religion: Sunni Muslim 64%
Shi'ite Muslim 28%
Currency: Turkish Lira = 100 kurus
Annual Income per person: $1,820
Annual Trade per person: $586
Adult Literacy: 81%
Life Expectancy (F): 68
Life Expectancy (M): 65

UKRAINE

Facts and Figures

Capital: Kiev 2,616,000 (1990)
Area (sq km): 603,700
Population: 52,100,000 (1992)
Language: Ukrainian, Russian
Religion: Ukrainian Orthodox,
Roman Catholic
Currency: Rouble = 100 kopeks
Annual Income per person: $2,340
Annual Trade per person: $600
Adult Literacy: Data not available
Life Expectancy (F): 75
Life Expectancy (M): 66

YUGOSLAVIA pre 1991

Facts and Figures

Capital: Belgrade 1,168,000 (1991)
Area (sq km): 102,000
Population: 10,460,000(1992)
Language: Serbo-Croat
Religion: Orthodox 65%, Muslim 19%
Currency: Dinar = 100 paras
Annual Income per person: $4,500
Annual Trade per person: $1,000
Adult Literacy: 89%
Life Expectancy (F): 75
Life Expectancy (M): 69

UNITED KINGDOM

Facts and Figures

Capital: London 6,679,700 (1991)
Area (sq km): 242,516
Population: 58,000,000
Language: English, Welsh, Scots-Gaelic
Religion: Anglican 57%
Roman Catholic 13%
Currency: Pound Sterling =100 pence
Annual Income per person: $16,080
Annual Trade per person: $7,140
Adult Literacy: 99%
Life Expectancy (F): 79
Life Expectancy (M): 73

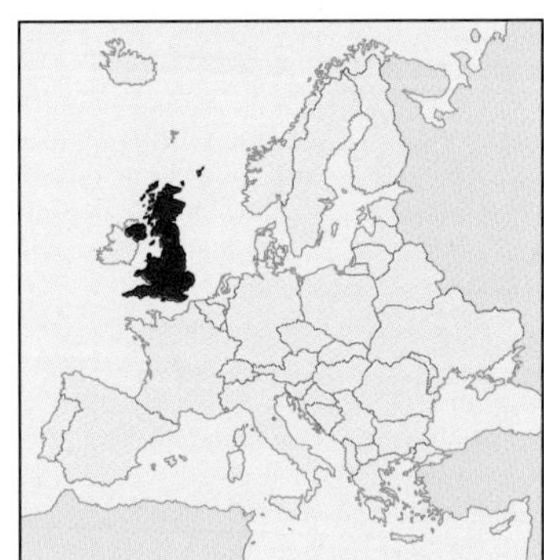

ENGLAND

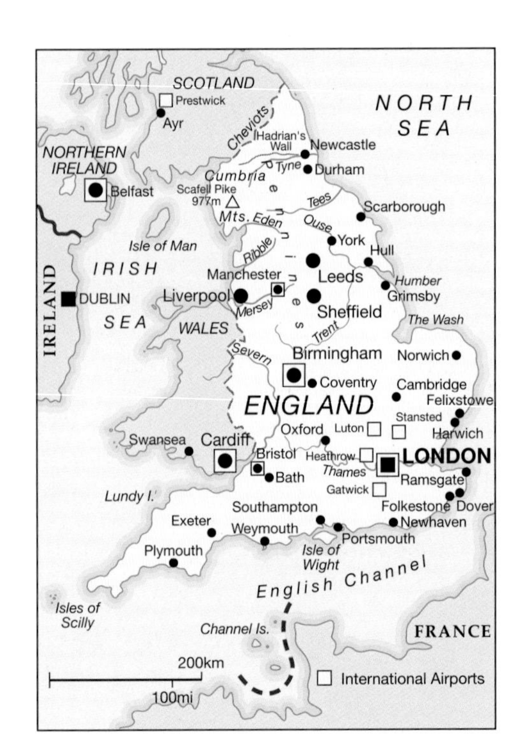

Facts and Figures

Capital: London 6,679,700 (1991)
Area (sq km): 130,439
Population: 47,536,000
Language: English
Religion: Protestant, Roman Catholic, Judaism, Islam
Currency: Pound Sterling = 100 pence
Annual Income per person:
Annual Trade per person:
Adult Literacy:
Life Expectancy (F):
Life Expectancy (M):

English Counties

1 NORTHUMBERLAND
2 TYNE & WEAR
3 DURHAM
4 CLEVELAND
5 CUMBRIA
6 LANCASHIRE
7 MERSEYSIDE
8 GREATER MANCHESTER
9 CHESHIRE
10 NORTH YORKSHIRE
11 WEST YORKSHIRE
12 SOUTH YORKSHIRE
13 HUMBERSIDE
14 LINCOLNSHIRE
15 NOTTINGHAMSHIRE
16 DERBYSHIRE
17 LEICESTERSHIRE
18 NORTHAMPTONSHIRE
19 CAMBRIDGESHIRE
20 NORFOLK
21 SUFFOLK
22 STAFFORDSHIRE
23 SHROPSHIRE
24 WEST MIDLANDS
25 WARWICKSHIRE
26 HEREFORD & WORCESTER
27 OXFORDSHIRE
28 BUCKINGHAMSHIRE
29 BEDFORDSHIRE
30 HERTFORDSHIRE
31 GREATER LONDON
32 ESSEX
33 KENT
34 EAST SUSSEX
35 WEST SUSSEX
36 SURREY
37 BERKSHIRE
38 HAMPSHIRE
39 ISLE OF WIGHT
40 GLOUCESTERSHIRE
41 AVON
42 WILTSHIRE
43 DORSET
44 SOMERSET
45 DEVON
46 CORNWALL & ISLES OF SCILLY

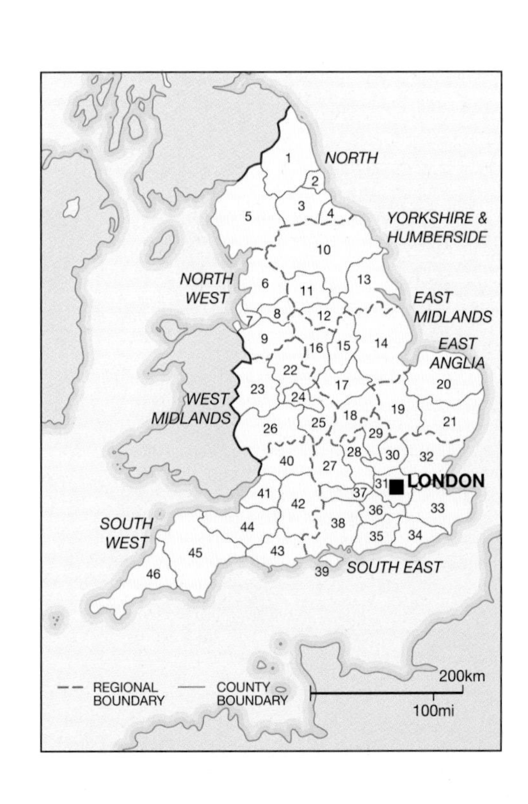

NORTHERN IRELAND

ATLANTIC OCEAN
Malin Head
Giant's Causeway
Rathlin I
SCOTLAND
North Channel
Lough Foyle
Coleraine
IRELAND
Derry
Bann
Strabane
Ballymena
Larne
NORTHERN IRELAND
Donegal
BELFAST
Lough Neagh
Lisburn
Donegal Bay
Lower Lough Erne
Enniskillen
Upper Lough Erne
Armagh
Strangford Lough
Newry
International Airports
100km
50mi
Dundalk
Carlingford Lough

Facts and Figures

Capital: Belfast 300,000
Area (sq km): 14,121
Population: 1,578,000
Language: English
Religion: Protestant, Roman Catholic
Currency: Pound Sterling = 100 pence
Annual Income per person:
Annual Trade per person:
Adult Literacy:
Life Expectancy (F):
Life Expectancy (M):

N. I. Counties

WALES

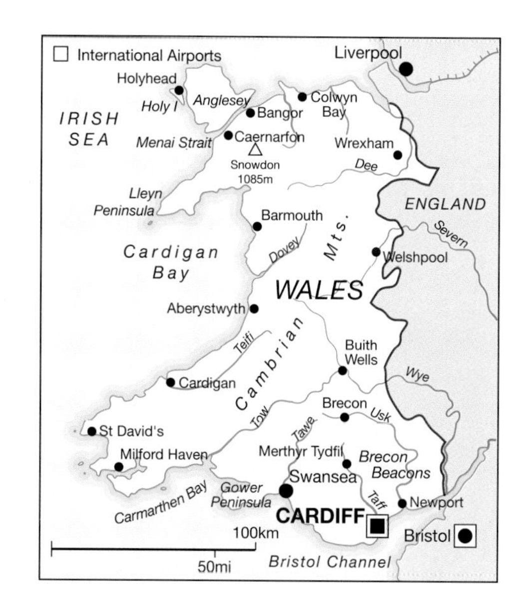

Facts and Figures

Capital: Cardiff 284,000
Area (sq km): 20,768
Population: 2,857,000
Language: English, Welsh
Religion: Protestant, Roman Catholic
Currency: Pound Sterling = 100 pence
Annual Income per person:
Annual Trade per person:
Adult Literacy:
Life Expectancy (F):
Life Expectancy (M):

100km
50mi
CLWYD
GWYNEDD
POWYS
DYFED
WEST GLAMORGAN
MID GLAMORGAN
GWENT
SOUTH GLAMORGAN
CARDIFF

Welsh Counties

SCOTLAND

Facts and Figures

Capital: Edinburgh 433,000
Area (sq km): 78,772
Population: 5,094,000
Language: English
Religion: Protestant, Roman Catholic
Currency: Pound Sterling = 100 pence
Annual Income per person:
Annual Trade per person:
Adult Literacy:
Life Expectancy (F):
Life Expectancy (M):

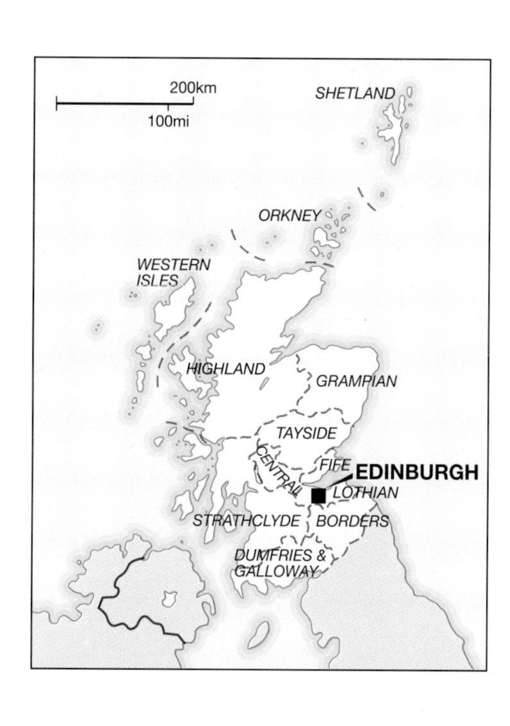

Scottish Regions

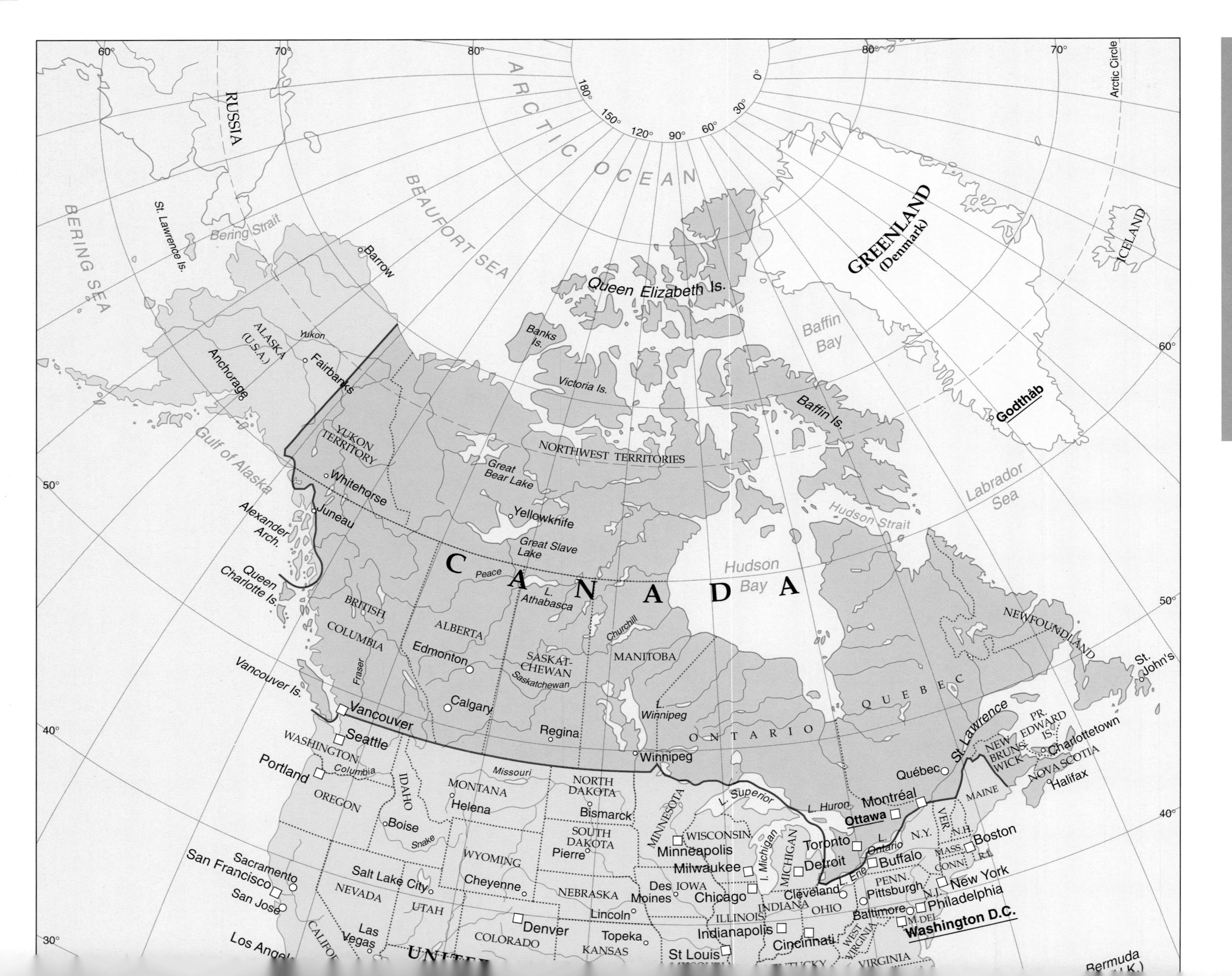
ARCTIC OCEAN
BEAUFORT SEA
BERING SEA
Labrador Sea
Baffin Bay
Hudson Bay
Hudson Strait
Bering Strait
Gulf of Alaska
Arctic Circle
ICELAND
GREENLAND
(Denmark)
Godthåb
RUSSIA
St. Lawrence Is.
ALASKA
(U.S.A.)
Barrow
Fairbanks
Anchorage
Yukon
Juneau
Alexander Arch.
Queen Charlotte Is.
Vancouver Is.
CANADA
YUKON TERRITORY
Whitehorse
NORTHWEST TERRITORIES
Yellowknife
Great Bear Lake
Great Slave Lake
Banks Is.
Victoria Is.
Queen Elizabeth Is.
Baffin Is.
BRITISH COLUMBIA
Fraser
Vancouver
ALBERTA
Edmonton
Calgary
Peace
L. Athabasca
SASKATCHEWAN
Saskatchewan
Regina
MANITOBA
Churchill
L. Winnipeg
Winnipeg
ONTARIO
QUEBEC
Québec
Montréal
Ottawa
Toronto
St. Lawrence
NEWFOUNDLAND
St. John's
NEW BRUNSWICK
PR. EDWARD IS.
Charlottetown
NOVA SCOTIA
Halifax
L. Superior
L. Huron
L. Michigan
L. Ontario
L. Erie
WASHINGTON
Seattle
OREGON
Portland
Columbia
IDAHO
Boise
Snake
MONTANA
Helena
Missouri
NORTH DAKOTA
Bismarck
SOUTH DAKOTA
Pierre
WYOMING
Cheyenne
NEBRASKA
Lincoln
COLORADO
Denver
KANSAS
Topeka
UTAH
Salt Lake City
NEVADA
Las Vegas
Sacramento
San Francisco
San Jose
Los Angel
MINNESOTA
Minneapolis
IOWA
Des Moines
WISCONSIN
Milwaukee
ILLINOIS
Chicago
St Louis
MICHIGAN
Detroit
INDIANA
Indianapolis
OHIO
Cleveland
Cincinnati
PENN.
Pittsburgh
Philadelphia
Baltimore
Washington D.C.
N.Y.
New York
Buffalo
VER.
N.H.
MAINE
MASS.
Boston
CONN.
R.I.
N.J.
M.DEL.
WEST VIRGINIA
VIRGINIA
Bermuda
180°
150°
120°
90°
60°
30°
0°
70°
60°
50°
40°
30°
80°

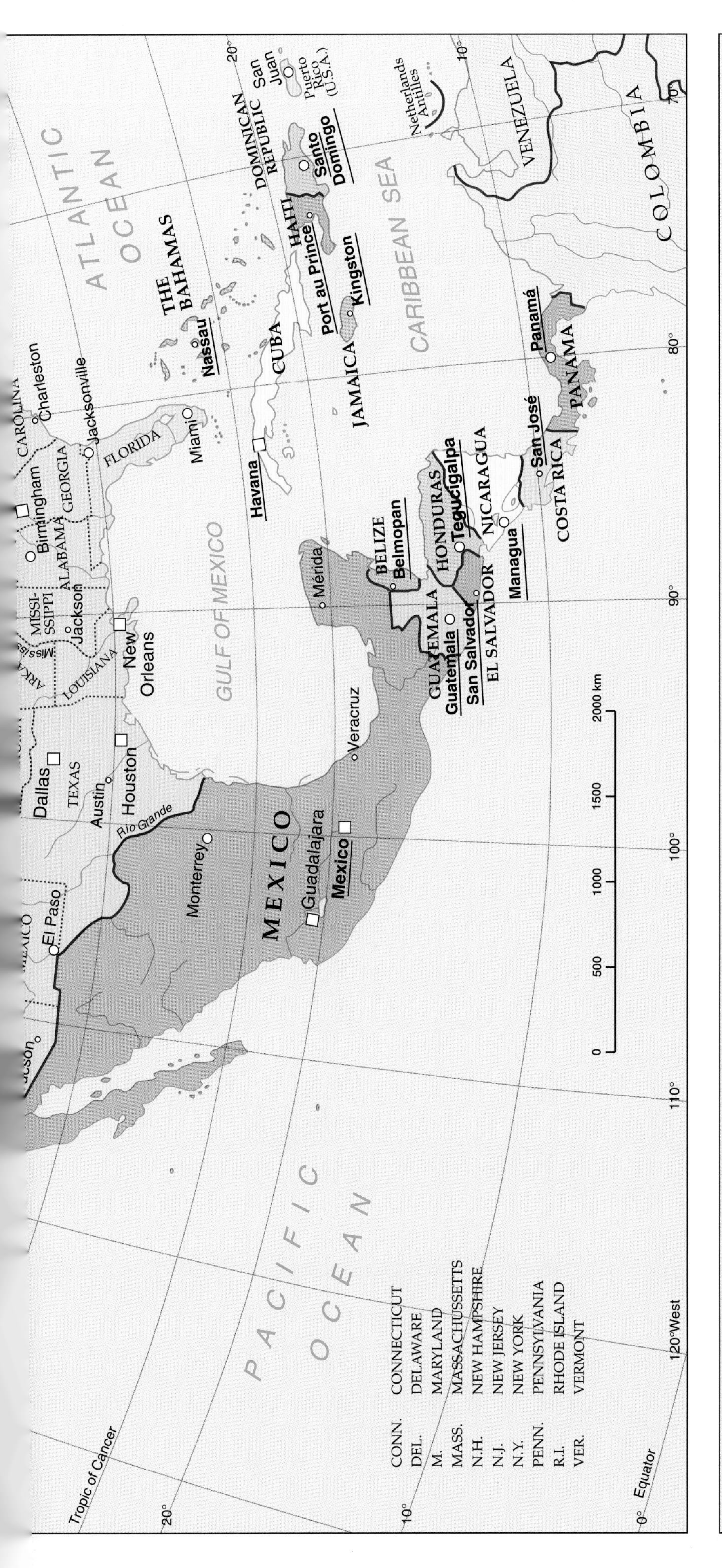
ATLANTIC OCEAN
THE BAHAMAS
Nassau
CUBA
Havana
HAITI
Port au Prince
DOMINICAN REPUBLIC
Santo Domingo
San Juan
Puerto Rico (U.S.A.)
Netherlands Antilles
JAMAICA
Kingston
CARIBBEAN SEA
VENEZUELA
COLOMBIA
PANAMA
Panamá
COSTA RICA
San José
NICARAGUA
Managua
HONDURAS
Tegucigalpa
EL SALVADOR
San Salvador
GUATEMALA
Guatemala
BELIZE
Belmopan
Mérida
GULF OF MEXICO
MEXICO
Mexico
Veracruz
Guadalajara
Monterrey
Rio Grande
El Paso
Austin
Houston
Dallas
TEXAS
LOUISIANA
New Orleans
MISSISSIPPI
Jackson
ALABAMA
Birmingham
GEORGIA
FLORIDA
Jacksonville
Miami
Charleston
CAROLINA
PACIFIC OCEAN
Tropic of Cancer
Equator
120°West
110°
100°
90°
80°
70°
20°
10°
0°
0
500
1000
1500
2000 km
CONN. CONNECTICUT
DEL. DELAWARE
M. MARYLAND
MASS. MASSACHUSETTS
N.H. NEW HAMPSHIRE
N.J. NEW JERSEY
N.Y. NEW YORK
PENN. PENNSYLVANIA
R.I. RHODE ISLAND
VER. VERMONT

COUNTRY	PAGE
Antigua & Barbuda	34
Bahamas	34
Barbados	34
Bermuda	35
Belize	35
Canada	36
Costa Rica	35
Cuba	36
Dominica	37
Dominican Republic	37
El Salvador	37
Greenland	37
Grenada	38
Guatemala	38
Haiti	38
Honduras	38
Jamaica	39
Mexico	39
Nicaragua	39
Panama	39
Puerto Rico	40
St. Christopher-Nevis	40
St. Lucia	40
St. Vincent and the Grenadines	40
Trinidad & Tobago	41
United States	41

ANTIGUA & BARBUDA

Facts and Figures

Capital: St. John's 30,000 (1982)
Area (sq km): 442
Population: 77,000 (1991)
Language: English
Religion: Protestant
Currency: Eastern Caribbean Dollar
Annual Income per person: $4,770
Annual Trade per person: N/A
Adult Literacy: 96%
Life Expectancy (F): 74
Life Expectancy (M): 70

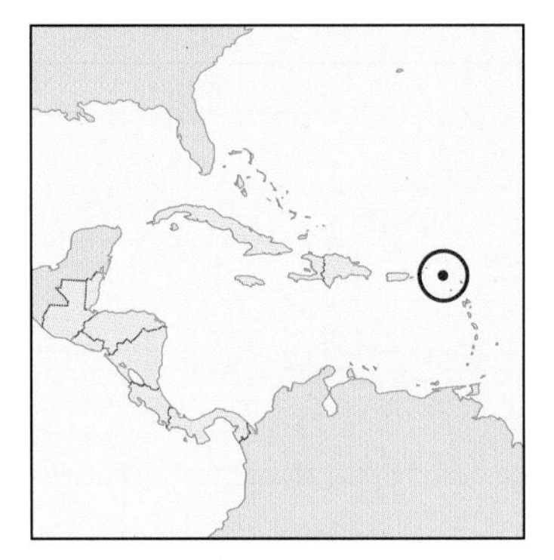

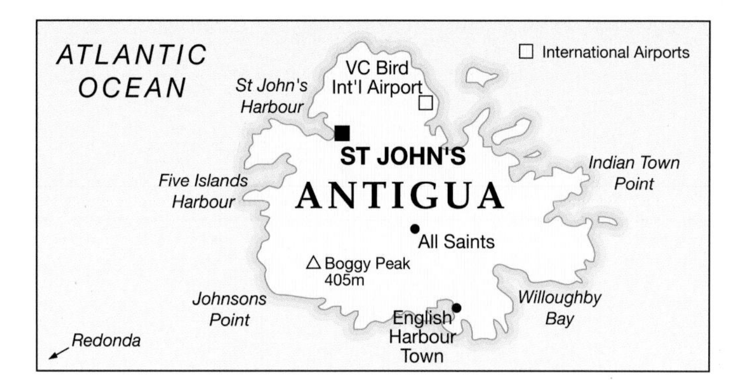

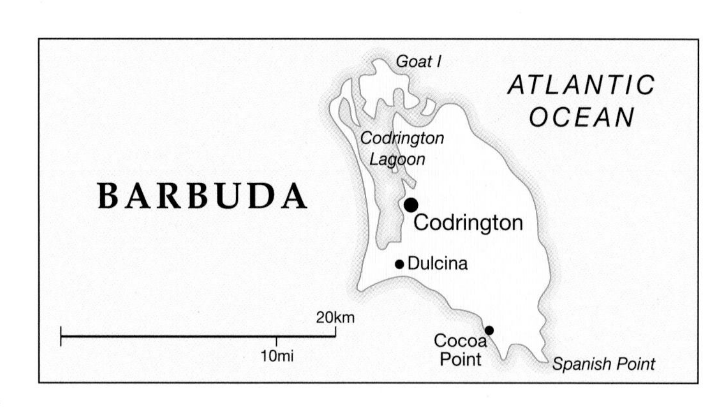

BAHAMAS

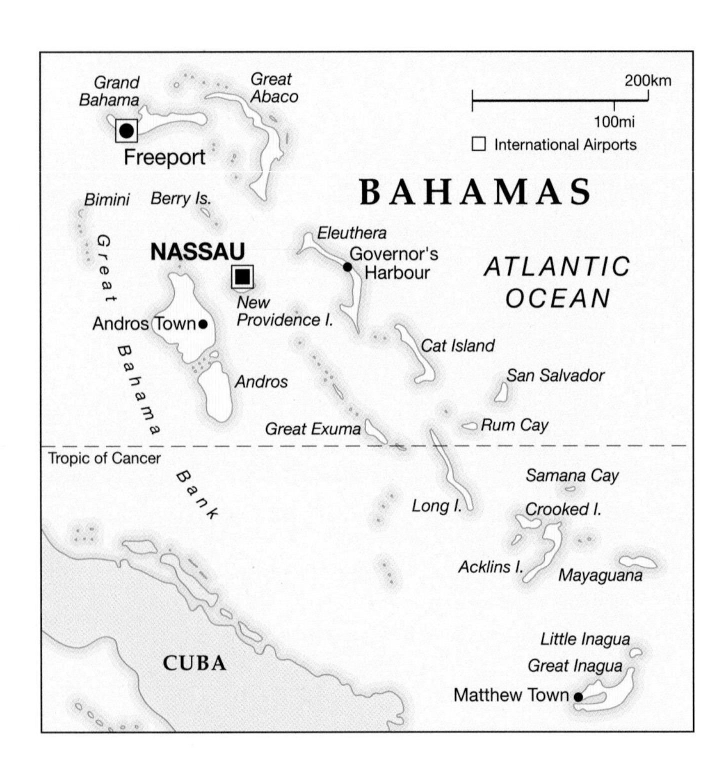

Facts and Figures

Capital: Nassau 172,000 (1990)
Area (sq km): 13,864
Population: 264,000 (1992)
Language: English, Creole
Religion: Protestant 69%
Roman Catholic 26%
Currency: Bahamian Dollar
Annual Income per person: $11,720
Annual Trade per person: $21,531
Adult Literacy: 99%
Life Expectancy (F): 76
Life Expectancy (M): 69

BARBADOS

Facts and Figures

Capital: Bridgetown 6,720 (1990)
Area (sq km): 430
Population: 258,600 (1991)
Language: English, Creole
Religion: Protestant 56%
Roman Catholic 4%
Currency: Barbados Dollar
Annual Income per person: $6,630
Annual Trade per person: $3,450
Adult Literacy: 99%
Life Expectancy (F): 78
Life Expectancy (M): 73

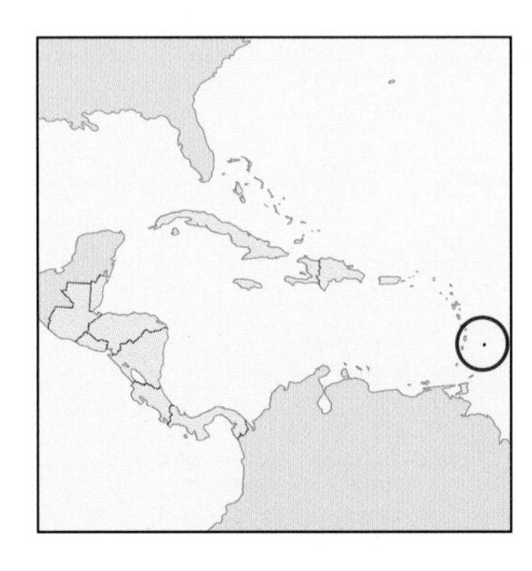

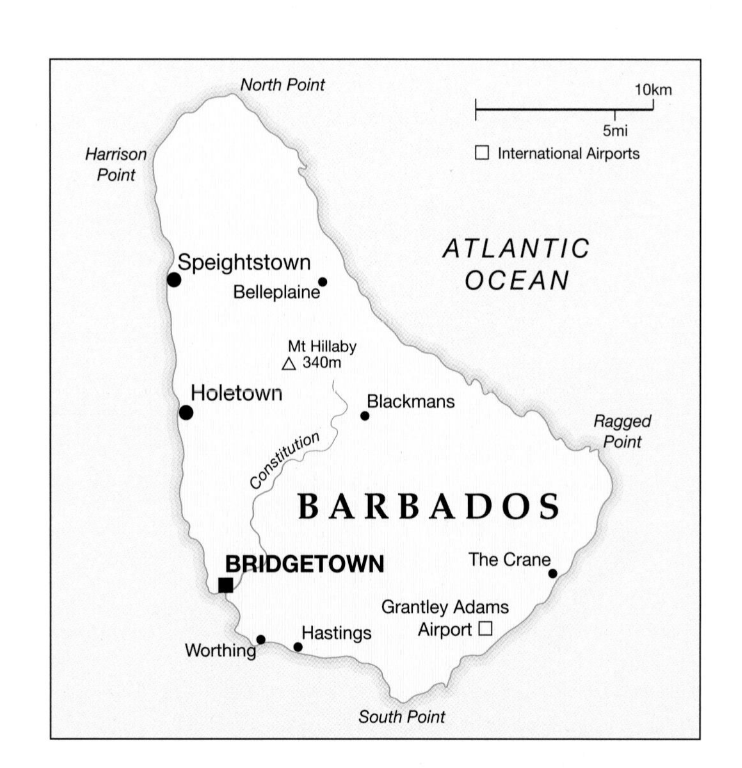

BELIZE

Yucatan Peninsula
MEXICO
Corozal
Orange Walk
Ambergris Cay
BELIZE
CARIBBEAN SEA
Belize City
Turneffe Is.
Belize
BELMOPAN
Lighthouse Reef
Reef
San Ignacio
Dangriga
Gulf of Honduras
GUATEMALA
Caracol
Maya Mts.
Glovers Reef
Barrier
Independence
San Antonio
Punta Gorda
International Airports
100km
50mi

Facts and Figures

Capital: Belmopan 5,276 (1990)
Area (sq km): 22,960
Population: 230,000 (1993)
Language: English, Creole, Spanish
Religion: Roman Catholic 62%
Protestant 28%
Currency: Belize Dollar
Annual Income per person: $2,050
Annual Trade per person: $1,953
Adult Literacy: 95%
Life Expectancy (F): 72
Life Expectancy (M): 67

BERMUDA

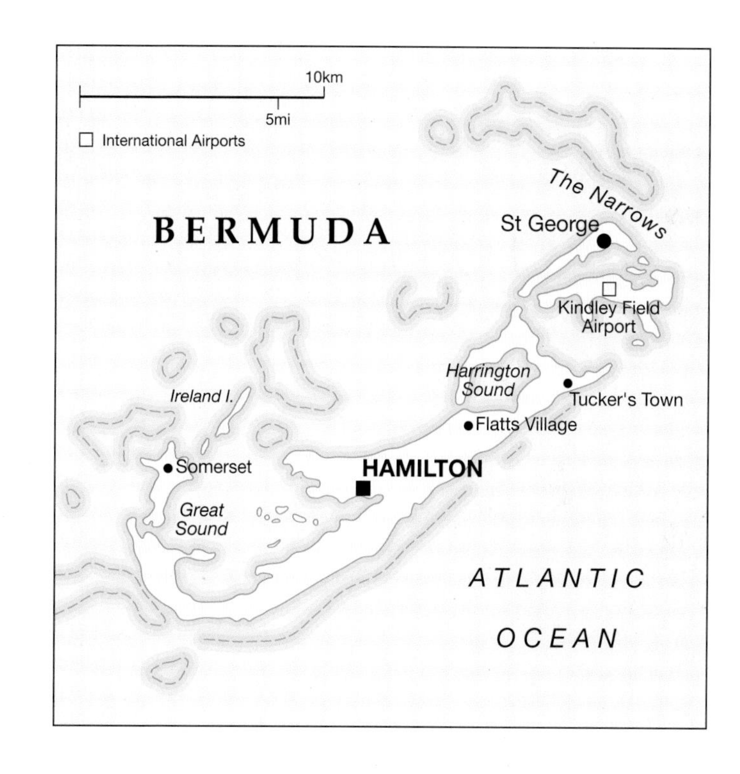

Facts and Figures

Capital: Hamilton 3,440 (1990)
Area (sq km): 53
Population: 71,950 (1992)
Language: English
Religion: Protestant 62%
Roman Catholic 14%
Currency: Bermuda Dollar
Annual Income per person: $24,000
Annual Trade per person: $9,750
Adult Literacy: 98%
Life Expectancy (F): 78
Life Expectancy (M): 72

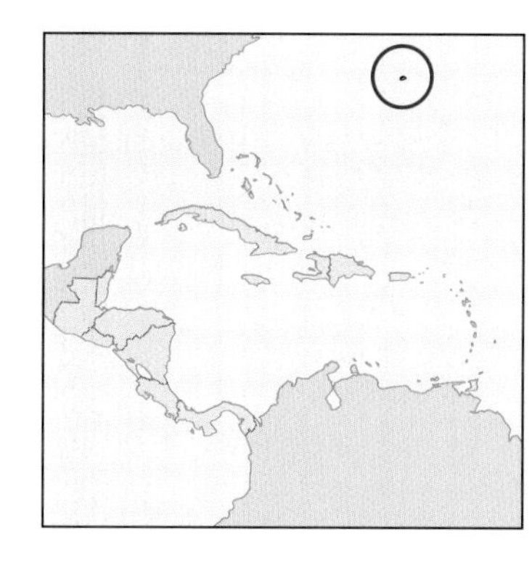

COSTA RICA

NICARAGUA
Lake Nicaragua
CARIBBEAN SEA
San Juan
Pan-American Highway
Liberia
COSTA RICA
Nicoya
Heredia
Puntarenas
Alajuela
Turrialba
Limón
SAN JOSÉ
Cartago
Nicoya Peninsula
Chirripó Grande 3819m
PANAMA
PACIFIC OCEAN
Rincón
Osa Peninsula
100km
50mi
International Airports
Point Burica

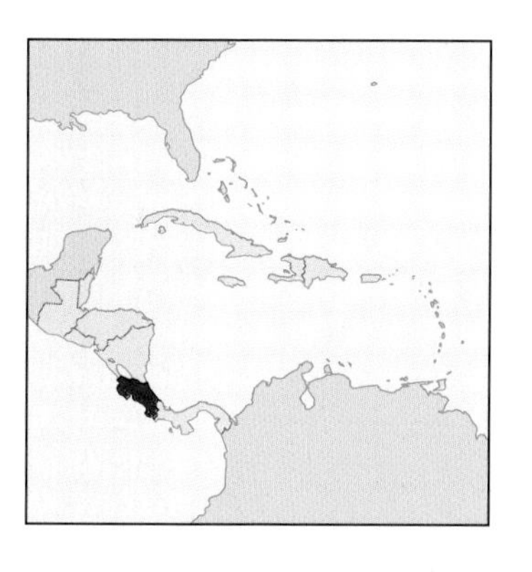

Facts and Figures

Capital: San José 245,000 (1984)
Area (sq km): 51,100
Population: 3,300,000 (1991)
Language: Spanish,Creole
Religion: Roman Catholic 90%
Currency: Colon=100 céntimos
Annual Income per person: $1,930
Annual Trade per person: $1,110
Adult Literacy: 93%
Life Expectancy (F): 78
Life Expectancy (M): 73

CANADA

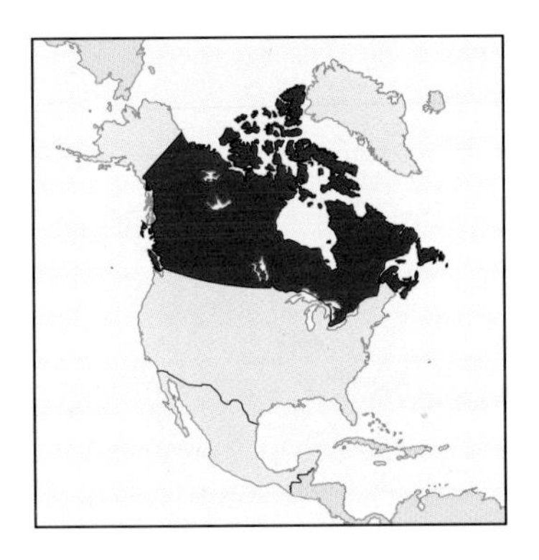

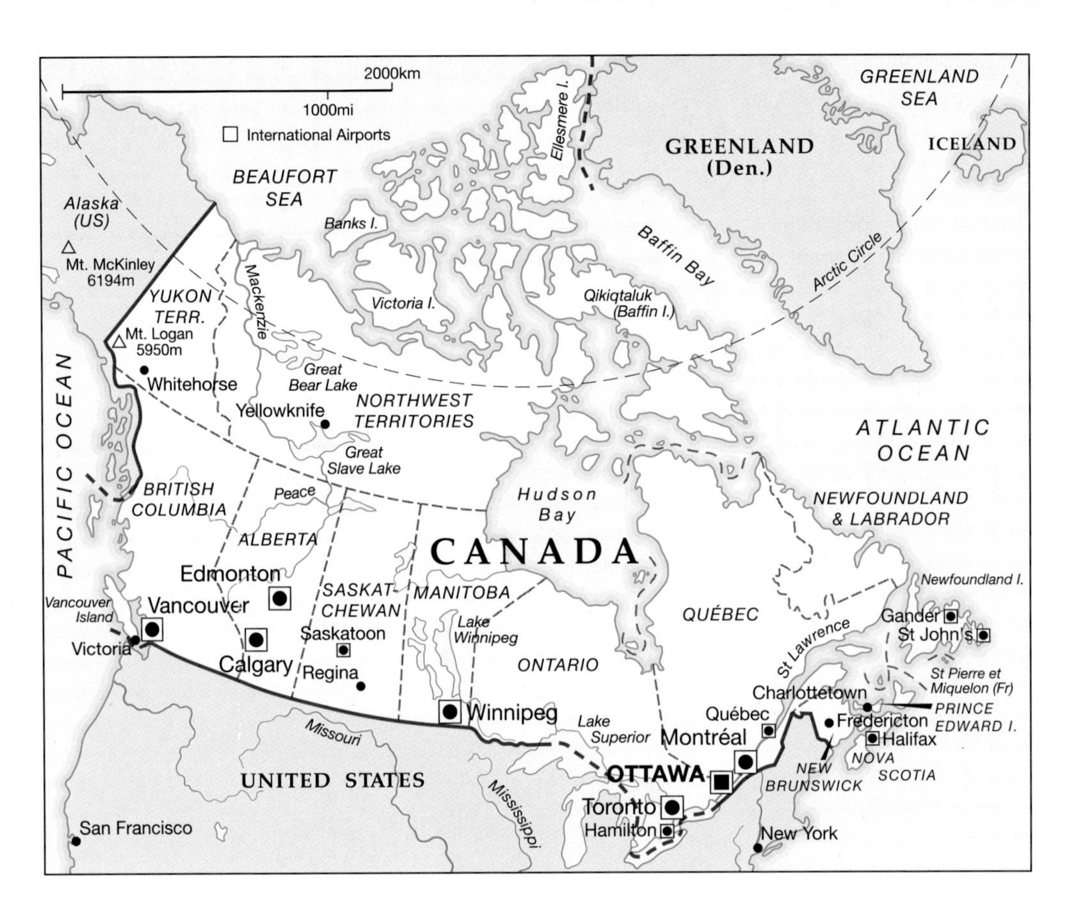

Facts and Figures

Capital: Ottawa 314,000 (1991)
Area (sq km): 9,970,610
Population: 27,400,000 (1992)
Language: English, French
Religion: Roman Catholic 47%
Protestant 41%
Currency: Canadian Dollar
Annual Income per person: $21,260
Annual Trade per person: $9,355
Adult Literacy: 99%
Life Expectancy (F): 81
Life Expectancy (M): 74

CUBA

Facts and Figures

Capital: Havana 2,014,800 (1986)
Area (sq km): 110,860
Population: 10,700,000 (1991)
Language: Spanish
Religion: Roman Catholic 39%
Protestant 3%
Currency: Cuban peso =100 centavos
Annual Income per person: $1,000
Annual Trade per person: $678
Adult Literacy: 75%
Life Expectancy (F): 78
Life Expectancy (M): 74

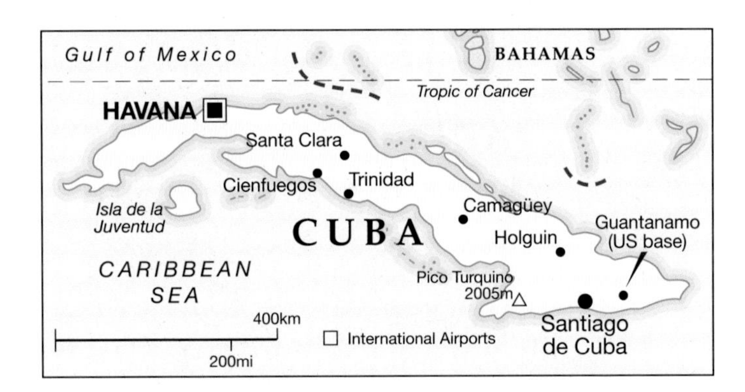

DOMINICA

Facts and Figures

Capital: Roseau 20,000 (1991)
Area (sq km): 751
Population: 108,800 (1991)
Language: English, French patois
Religion: Roman Catholic 80%
Protestant 15%
Currency: East Caribbean Dollar
= 100 cents
Annual Income per person: $2,440
Annual Trade per person: $2,163
Adult Literacy: 97%
Life Expectancy (F): 79
Life Expectancy (M): 73

DOMINICAN REPUBLIC
Puerto Plata
Yaque del Norte
Cabo Francés Viejo
Santiago
Samaná Peninsula
HAITI
Cordillera Central
La Vega
San Francisco de Macorís
Pico Duarte 3175 m
San Juan
Lago de Enriquillo
Yaque del Sur
SANTO DOMINGO
Cabo Engaño
La Romana
San Pedro de Macorís
Barahona
Isla Saona
CARIBBEAN SEA
Isla Beata
Cabo Beata
International Airports
100km
50mi

DOMINICAN REPUBLIC

Facts and Figures

Capital: Santo Domingo 1,601,000 (1986)
Area (sq km): 48,442
Population: 7,310,000 (1991)
Language: Spanish
Religion: Roman Catholic 92%
Currency: Peso = 100 centavos
Annual Income per person: $950
Annual Trade per person: $324
Adult Literacy: 83%
Life Expectancy (F): 70
Life Expectancy (M): 65

EL SALVADOR

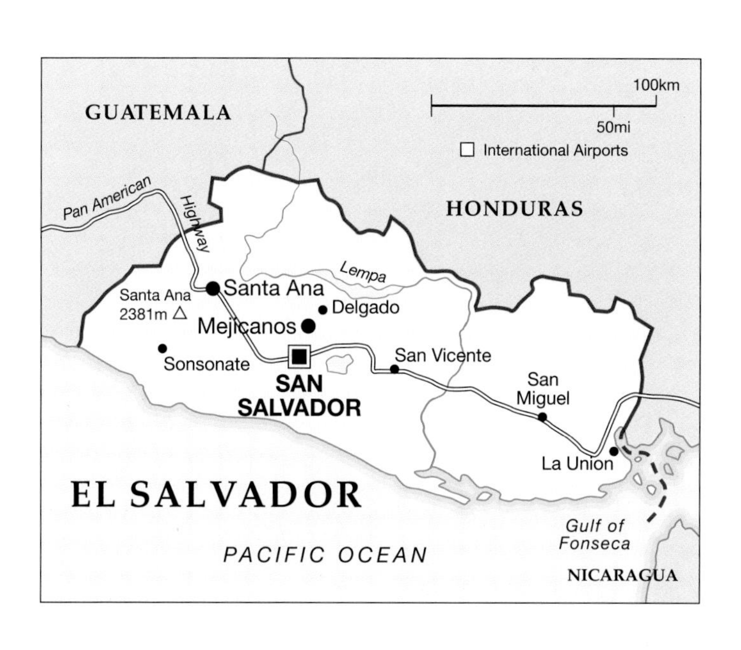

Facts and Figures

Capital: San Salvador 1,522,130 (1992)
Area (sq km): 21,041
Population: 5,050,000 (1992)
Language: Spanish
Religion: Roman Catholic 91%
Protestant 4%
Currency: Colón = 100 centavos
Annual Income per person: $1,070
Annual Trade per person: $250
Adult Literacy: 73%
Life Expectancy (F): 69
Life Expectancy (M): 64

GREENLAND

Facts and Figures

Capital: Godthåb (Nuuk) 12,200 (1993)
Area (sq km): 2,175,600
Population: 55,000 (1993)
Language: Eskimo dialects, Danish
Religion: Evanjelical Lutherans 98%
Currency: Danish Krone = 100 ore
Annual Income per person: $6,000
Annual Trade per person: $15,017
Adult Literacy: Data not available
Life Expectancy (F): 69
Life Expectancy (M): 63

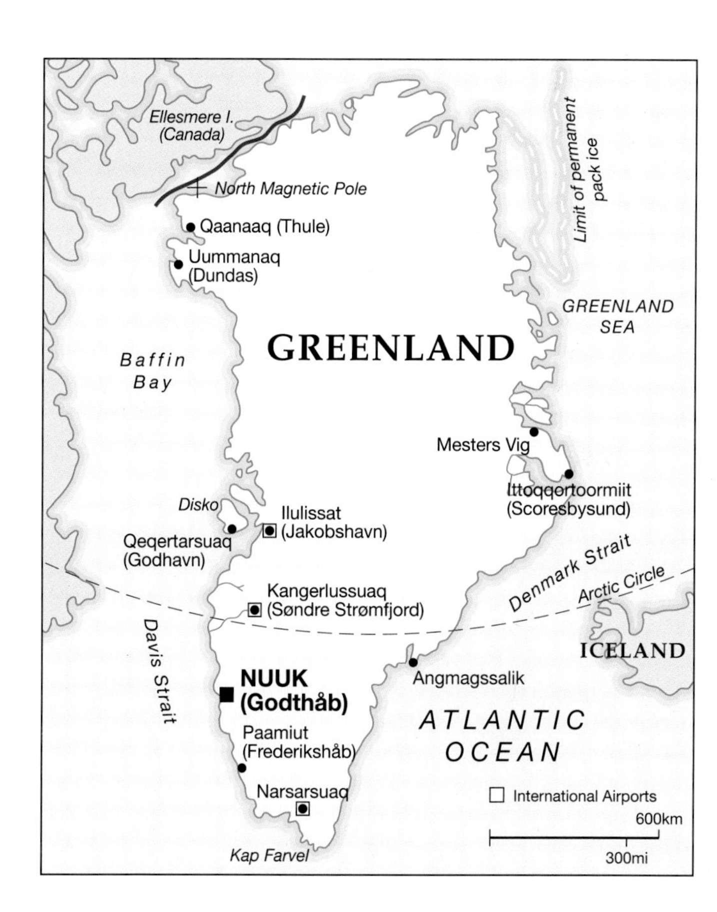

GRENADA

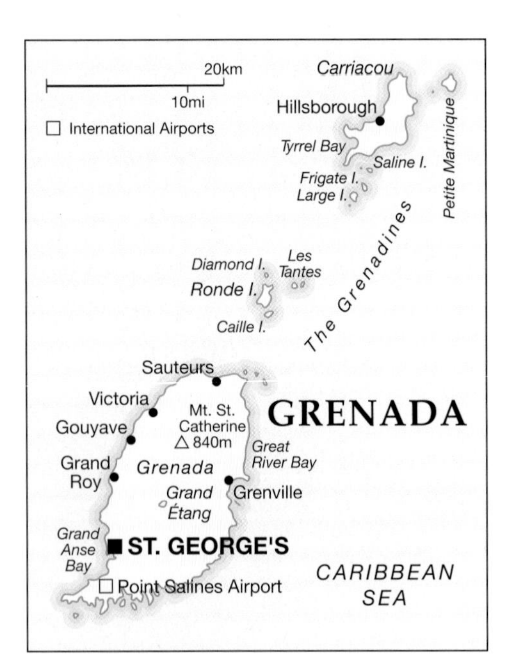

Facts and Figures

Capital: St George's 35,750 (1980)
Area (sq km): 344
Population: 95,350 (1993)
Language: English, French patois
Religion: Roman Catholic 53% Anglican 14%
Seventh Day Adventist 9%
Pentecostal 7%
Currency: Eastern Caribbean $
Annual Income per person: $2,180
Annual Trade per person: $1,342
Adult Literacy: 96%
Life Expectancy (F): 74
Life Expectancy (M): 69

GUATEMALA

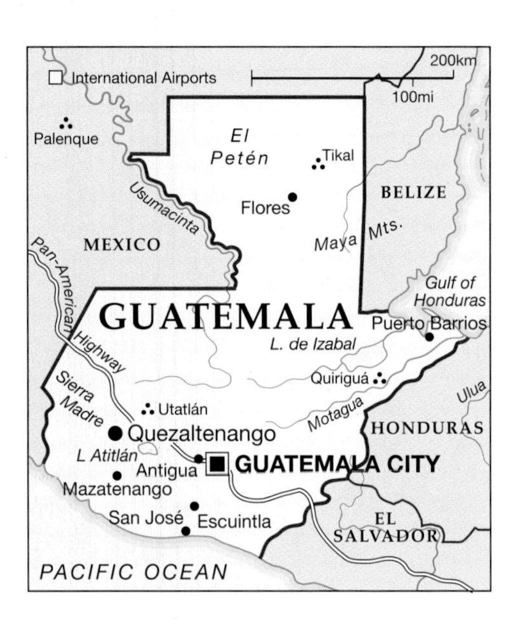

Facts and Figures

Capital: Guatemala City 2,000,000 (1989)
Area (sq km): 108,890
Population: 9,740,000 (1992)
Language: Spanish, Indian dialects
Religion: Roman Catholic 71%
Pentecostal 23%
Other Protestant 7%
Currency: Quetzal = 100 centavos
Annual Income per person: $930
Annual Trade per person: $286
Adult Literacy: 55%
Life Expectancy (F): 67
Life Expectancy (M): 62

HAITI

Facts and Figures

Capital: Port-au-Prince 1,402,000 (1991)
Area (sq km): 27,750
Population: 6,760,000 (1992)
Language: Haitian Créole,French
Religion: Roman Catholic 75% Other Christian 10% Voodoo
Currency: Gourde = 100 centimes
Annual Income per person: $370
Annual Trade per person: $70
Adult Literacy: 53%
Life Expectancy (F): 58
Life Expectancy (M): 55

HONDURAS

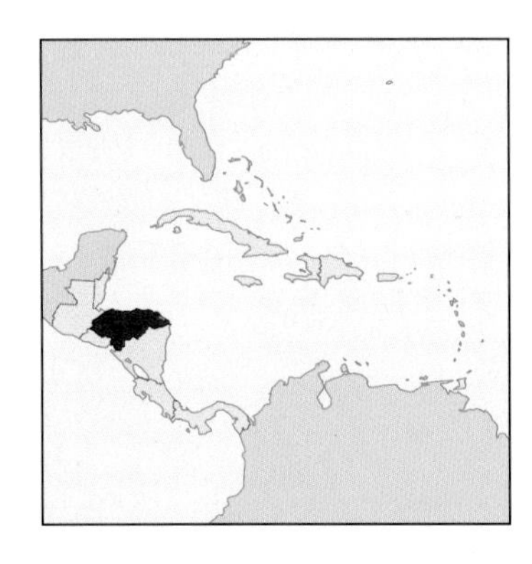

Facts and Figures

Capital: Tegucigalpa 678,700 (1988)
Area (sq km): 112,088
Population: 5,260,000
Language: Spanish
Religion: Roman Catholic 97%
Currency: Lempira = 100 centavos
Annual Income per person: $570
Annual Trade per person: $321
Adult Literacy: 73%
Life Expectancy (F): 68
Life Expectancy (M): 64

JAMAICA

Facts and Figures

Capital: Kingston 587.800 (1991)
Area (sq km): 11,425
Population: 2,450,000 (1992)
Language: English, English Creole
Religion: Protestant 70%
Roman Catholic 8%
Currency: Jamaican Dollar $
Annual Income per person: $1,380
Annual Trade per person: $1,242
Adult Literacy: 98%
Life Expectancy (F): 76
Life Expectancy (M): 71

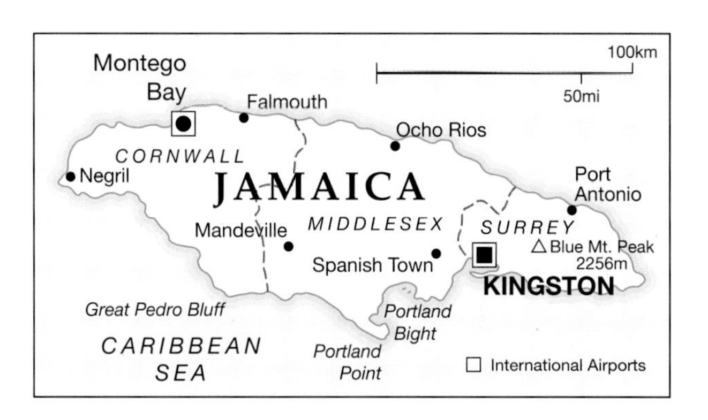

MEXICO

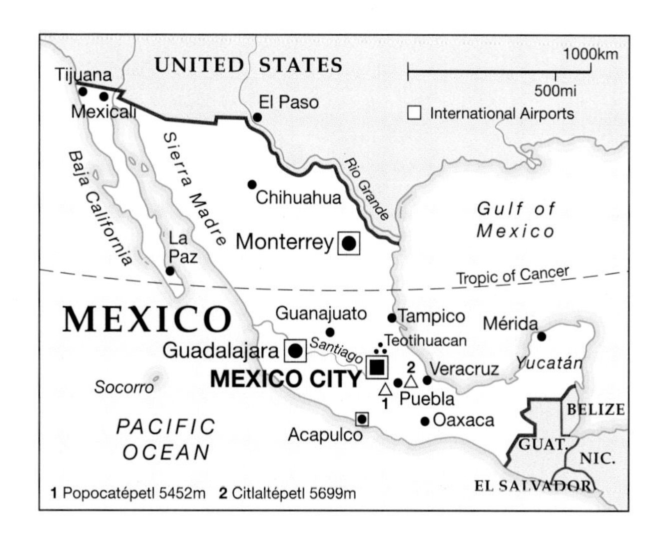

Facts and Figures

Capital: Mexico City 15,047,770 (1990)
Area (sq km): 1,967,200
Population: 91,600,000 (1992)
Language: Spanish
Religion: Roman Catholic 94%
Protestant 3%
Currency: Peso = 100 centavos
Annual Income per person: $2,870
Annual Trade per person: $743
Adult Literacy: 87%
Life Expectancy (F): 74
Life Expectancy (M): 67

NICARAGUA

Facts and Figures

Capital: Managua 682,100 (1985)
Area (sq km): 130,682
Population: 4,200,000 (1991)
Language: Spanish
Religion: Roman Catholic 88%
Currency: Córdoba =100centavos
Annual Income per person: $340
Annual Trade per person: $349
Adult Literacy: 81%
Life Expectancy (F): 68
Life Expectancy (M): 65

PANAMA

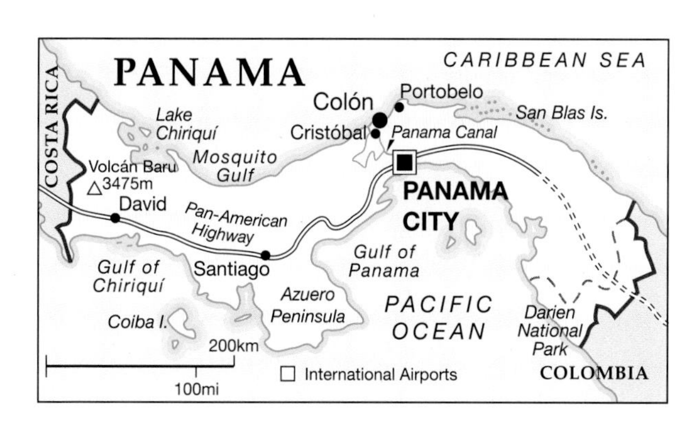

Facts and Figures

Capital: Panama City 584,800 (1990)
Area (sq km): 77,080
Population: 2,330,000 (1990)
Language: Spanish, English
Religion: Roman Catholic 85%
Protestant 5% Moslem 5%
Currency: Balboa = 100centesimos
Annual Income per person: $2180
Annual Trade per person: $825
Adult Literacy: 72%
Life Expectancy (F): 75
Life Expectancy (M): 71

PUERTO RICO

Facts and Figures

Facts and Figures for Puerto Rico are incorporated into the details given for the United States of America

ST. CHRISTOPHER NEVIS

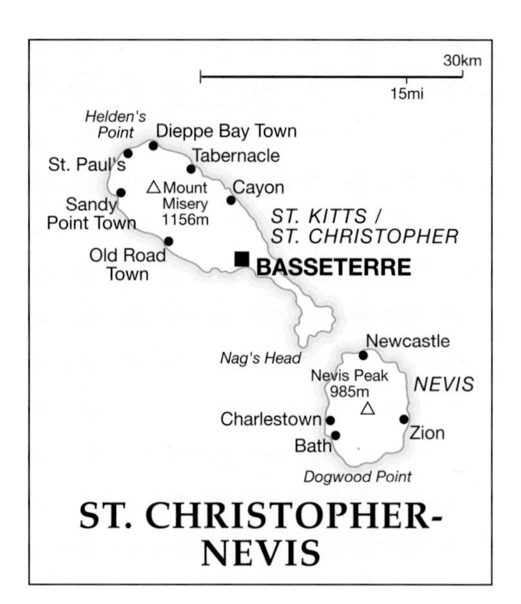

Facts and Figures

Capital: Basseterre 14,300 (1980)
Area (sq km): 262
Population: 40,620 (1991)
Language: English
Religion: Protestant 76%
Roman Catholic 11%
Currency: East Caribbean Dollar =100cents
Annual Income per person: $3,960
Annual Trade per person: $3,200
Adult Literacy: 92%
Life Expectancy (F): 71
Life Expectancy (M): 64

ST. LUCIA

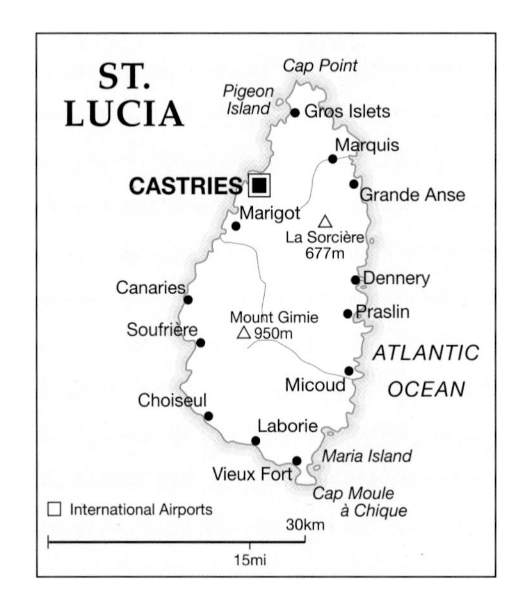

Facts and Figures

Capital: Castries 53,900 (1992)
Area (sq km): 617
Population: 136,000 (1992)
Language: English,French patois
Religion: Roman Catholic 82%
Protestant 10%
Currency: East Caribbean Dollar =100cents
Annual Income per person: $2,500
Annual Trade per person: $2,267
Adult Literacy: 93%
Life Expectancy (F): 74
Life Expectancy (M): 69

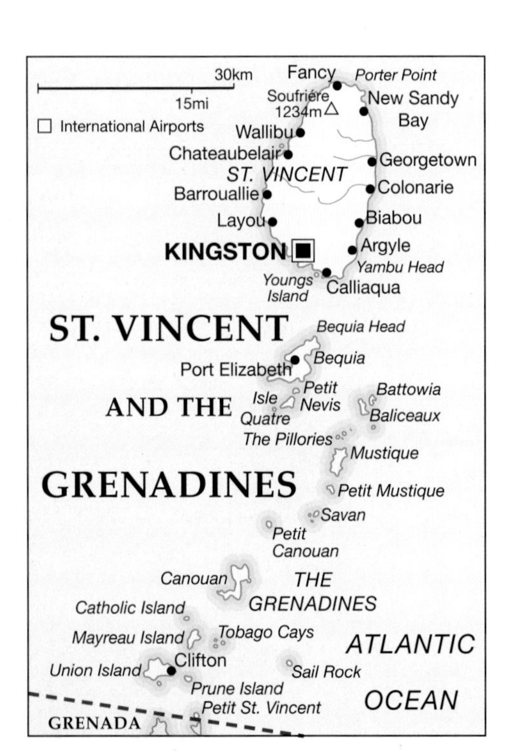

ST. VINCENT

Facts and Figures

Capital: Kingstown 26,600 (1991)
Area (sq km): 388
Population: 107,600 (1991)
Language: English,French patois
Religion: Protestant 53%
Roman Catholic 12%
Currency: E. Carib. Dollar =100 cents
Annual Income per person: $1,730
Annual Trade per person: $1,800
Adult Literacy: 84%
Life Expectancy (F): 72
Life Expectancy (M): 68

60km
30mi
□ International Airports

Charlotteville
Roxborough
Scarborough
TOBAGO

TRINIDAD & TOBAGO

PORT OF SPAIN
VENEZUELA
Dragon's Mouths
Toco
Galera Point
Mt. Aripo △940m
Caroni
Arima
Matura Bay
Piarco Airport
Sangre Grande
Cocos Bay
Gulf of Paria
TRINIDAD
CARIBBEAN SEA
San Fernando
Point Fortin
Pitch Lake
Ortoire
Mayaro Bay
Moruga
Galeota Point
Icacos Point

TRINIDAD & TOBAGO

Facts and Figures

Capital: Port-of-Spain 58,400 (1990)
Area (sq km): 5,124
Population: 1,250,000 (1991)
Language: English
Religion: Roman Catholic 29% Protestant 11% Hindu 24% Muslim 6%
Currency: Trinidad and Tobago $ =100 cents
Annual Income per person: $3,620
Annual Trade per person: $2,902
Adult Literacy: 96%
Life Expectancy (F): 75
Life Expectancy (M): 70

UNITED STATES OF AMERICA

Facts and Figures

Capital: Washington D. C. 606,900 (1990)
Area (sq km): 9,158,960
Population: 252,180,000 (1991)
Language: English, Spanish
Religion: Christian, Jewish
Currency: US Dollar = 100 cents
Annual Income per person: $22,560
Annual Trade per person: $3,922
Adult Literacy: 99%
Life Expectancy (F): 79
Life Expectancy (M): 72

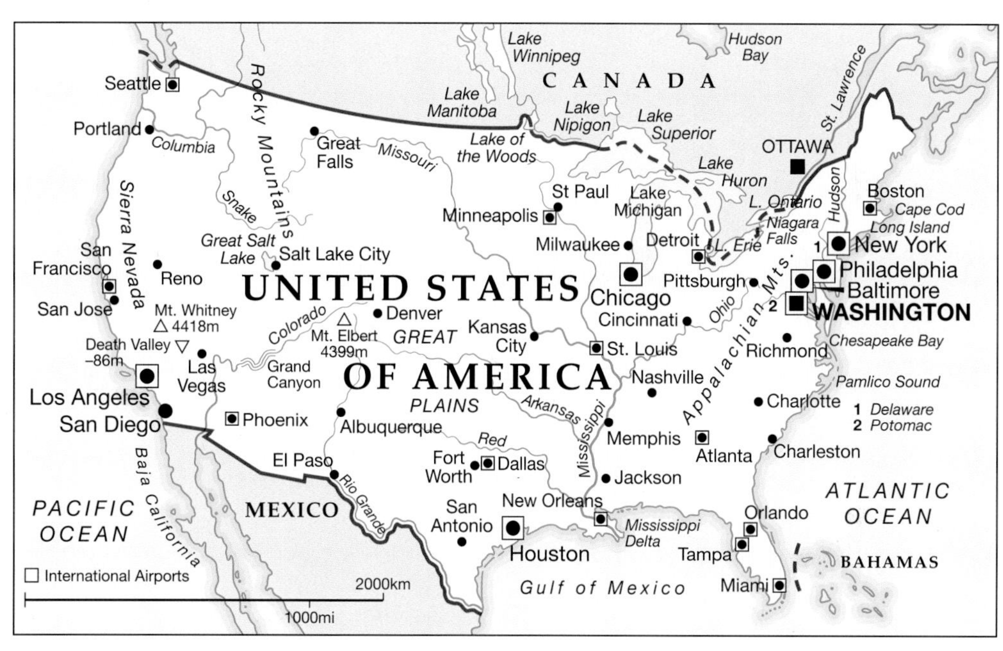

US States

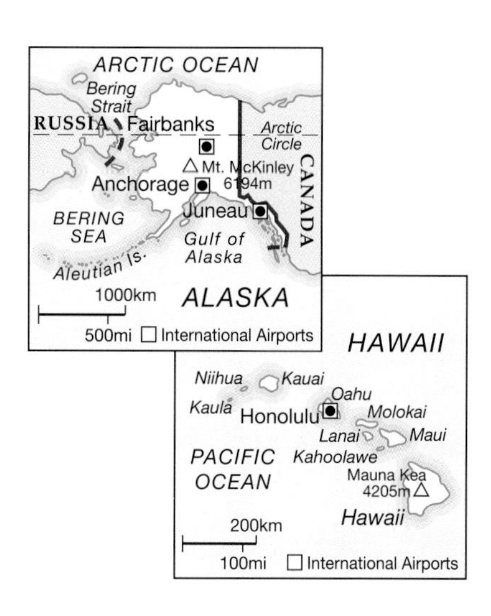

CARIBBEAN SEA
HONDURAS
NICARAGUA
COSTA RICA
PANAMA
Neth. Antilles
ANTIGUA & BARBUDA
Guadeloupe (Fr.)
DOMINICA
Martinique (Fr.)
ST. LUCIA
ST. VINCENT
BARBADOS
GRENADA
TRINIDAD & TOBAGO
Port of Spain
VENEZUELA
Caracas
Maracaibo
Barcelona
Güiria
Ciudad Guayana
GUYANA
Georgetown
SURINAM
Paramaribo
FRENCH GUIANA
Cayenne
COLOMBIA
Bogotá
Barranquilla
Cartagena
Montería
Medellín
Cali
ECUADOR
Quito
Guayaquil
Manta
Loja
Equator
PERU
Lima
Callao
Trujillo
Iquitos
Huancayo
Cuzco
Arequipa
L. Titicaca
Japurá
Ucayali
BOLIVIA
La Paz
Sucre
Oruro
Santa Cruz
Arica
Antofagasta
PARAGUAY
Asunción
Concepcíon
Paraguay
BRAZIL
Brasília
AMAZONAS
Amazonas
Manaus
Negro
Madeira
Madre de Dios
Humaitá
Pôrto Velho
Labrea
Rio Branco
Cruzeiro do Sul
ACRE
RONDÔNIA
RORAIMA
Boa Vista
PARÁ
Santarém
Xingu
AMAPÁ
Macapá
Belém
MARANHÃO
Imperatriz
Carolina
Sao Luis
Parnaíba
Teresina
PIAUÍ
CEARA
Fortaleza
RIO GRANDE DO NORTE
Natal
PARAIBA
PERNAMBUCO
Recife
ALAGOAS
Maceió
SERGIPE
Salvador
BAHIA
Juázeiro
São Francisco
TOCANTINS
Araguaia
GOIÁS
Goiânia
MATO GROSSO
Cuiabá
Corumbá
MATO GROSSO DE SUL
Campo Grande
MINAS GERAIS
Belo Horizonte
ESPIRITO SANTO
Vitória
RIO DE JANERIO
Rio de Janeiro
SÃO PAULO
São Paulo
Campinas
PARANÁ
Paraná
Curitiba
Foz do Iguacu
SANTA CATARINA
RIO GRANDE
CORRIENTES
MISIONES
FORMOSA
CHACO
SANTA
SALTA
Salta
TUCUMAN
SANTIAGO DEL ESTERO
CATAMARCA
Tropic of Capricorn
0
200
400
600
800
2000 km
10°
0°
10°
20°
40°
50°
60°
70°
80°

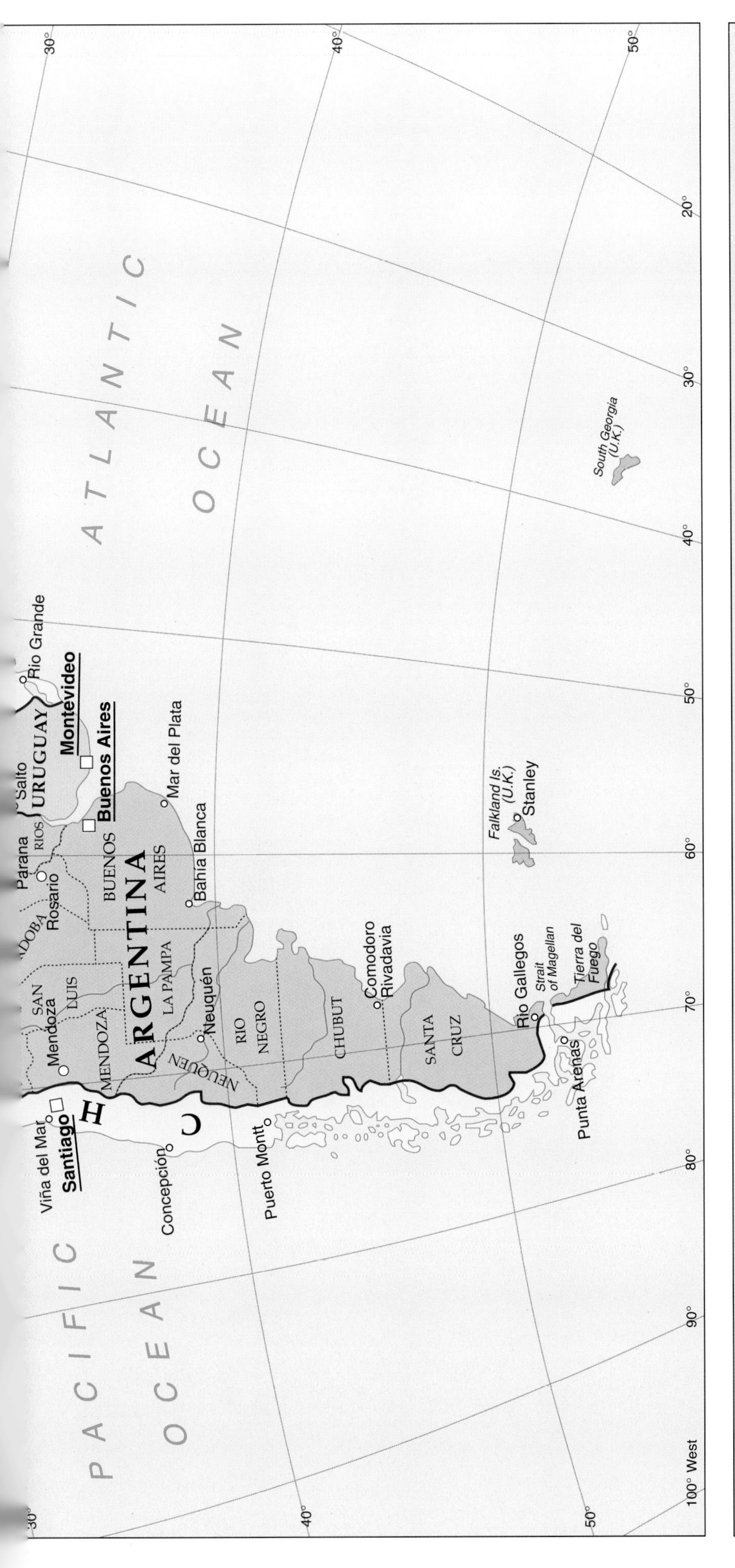

COUNTRY	PAGE
Argentina	44
Bolivia	44
Brazil	44
Chile	45
Colombia	45
Ecuador	45
Guyana	46
Paraguay	46
Peru	46
Surinam	47
Uruguay	47
Venezuela	47

ARGENTINA

Facts and Figures

Capital: Buenos Aires 9,927,400 (1980)
Area (sq km): 2,780,092
Population: 32,370,000 (1991)
Language: Spanish
Religion: Roman Catholic 94%
Protestant 2%
Currency: Peso
Annual Income per person: $2,780
Annual Trade per person: $613
Adult Literacy: 95%
Life Expectancy (F): 75
Life Expectancy (M): 68

BOLIVIA

Facts and Figures

Capital: Sucre 105,800 (1988)
Area (sq km): 1,098,580
Population: 7,610,000 (1991)
Language: Spanish, Aymara, Quechua
Religion: Roman Catholic 89%
Currency: Boliviano = 100 centavos
Annual Income per person: $650
Annual Trade per person: $237
Adult Literacy: 78%
Life Expectancy (F): 58
Life Expectancy (M): 54

BRAZIL

Facts and Figures

Capital: Brasilia 1,596,270 (1991)
Area (sq km): 8,511,996
Population: 159,100,000 (1991)
Language: Portuguese, Spanish, English
Religion: Roman Catholic 89%
Protestant 7%
Currency: Cruzeiro Real = 100 centavos
Annual Income per person: $2,920
Annual Trade per person: $343
Adult Literacy: 81%
Life Expectancy (F): 69
Life Expectancy (M): 64

PERU
BOLIVIA
BRAZIL
Paraguay
Arica
S
Atacama Desert
Tropic of Capricorn
PARAGUAY
Antofagasta
E
Ojos del Salado 6910m
Paraná
La Serena
D
Viña del mar
Aconcagua 6960m
Valparaíso
URUGUAY
SANTIAGO
N
BUENOS AIRES
Talcahuano
Concepción
ARGENTINA
CHILE
Valdivia
A
SOUTH ATLANTIC OCEAN
Chiloe I.
PACIFIC OCEAN
Patagonia
International Airports
1000km
500mi
Punta Arenas
Tierra del Fuego
Cape Horn

CHILE

Facts and Figures

Capital: Santiago 4,858,350 (1987)
Area (sq km): 736,900
Population: 13,440,000 (1993)
Language: Spanish
Religion: Roman Catholic 80%
Protestant 6%
Currency: Chilean Peso = 100 centavos
Annual Income per person: $2,160
Annual Trade per person: $1,221
Adult Literacy: 93%
Life Expectancy (F): 76
Life Expectancy (M): 69

COLOMBIA

Facts and Figures

Capital: Bogotá 4,921,300 (1992)
Area (sq km): 1,141,750
Population: 33,390,000 (1992)
Language: Spanish
Religion: Roman Catholic 94%
Currency: Colombian Peso =
100 centavos
Annual Income per person: $1,280
Annual Trade per person: $364
Adult Literacy: 87%
Life Expectancy (F): 72
Life Expectancy (M): 66

PACIFIC OCEAN
COLOMBIA
Esmeraldas
Tulcán
Caqueta
Putumayo
Equator
QUITO
Coca
COSTA
SIERRA
Cotopaxi 5896m
Napo
Manta
Portoviejo
Vinces
Chimborazo 6310m
ORIENTE
Riobamba
Sangay 5230m
Guayaquil
ECUADOR
Gulf of Guayaquil
Isla Puná
Cuenca
PERU
Machala
ANDES
Loja
International Airports
Cordillera del Condor
200km
100mi

ECUADOR

Facts and Figures

Capital: Quito 1,100,850 (1990)
Area (sq km): 272,045
Population: 9,650,000 (1990)
Language: Spanish, Quechua
Religion: Roman Catholic 91%
Protestant 6%
Currency: Sucre = 100 centavos
Annual Income per person: $1,020
Annual Trade per person: $474
Adult Literacy: 86%
Life Expectancy (F): 69
Life Expectancy (M): 65

GUYANA

Facts and Figures

Capital: Georgetown 188,000 (1983)
Area (sq km): 214,970
Population: 990,000 (1989)
Language: English, Hindi, Urdu
Religion: Protestant 34% Hindu 34% Roman Catholic 18% Muslim 9%
Currency: Guyanan Dollar = 100 cents
Annual Income per person: $290
Annual Trade per person: $959
Adult Literacy: 96%
Life Expectancy (F): 68
Life Expectancy (M): 62

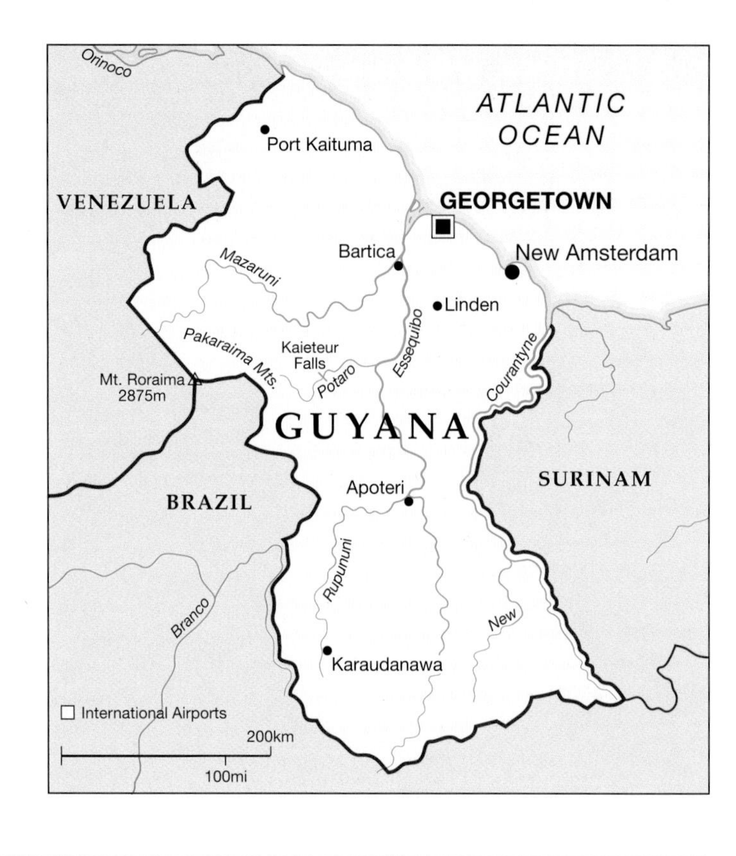

PARAGUAY

Facts and Figures

Capital: Asunción 607,700 (1990)
Area (sq km): 406,750
Population: 4,500,000 (1993)
Language: Spanish, Guaraní
Religion: Roman Catholic 96%
Currency: Guaraní = 100 céntimos
Annual Income per person: $1,210
Annual Trade per person: $447
Adult Literacy: 90%
Life Expectancy (F): 70
Life Expectancy (M): 65

PERU

Facts and Figures

Capital: Lima 5,759,700 (1993)
Area (sq km): 1,244,300
Population: 22,130,000 (1993)
Language: Spanish, Quechua, Aymara
Religion: Roman Catholic 93%
Currency: Nuevo sol = 100 centavos
Annual Income per person: $1,020
Annual Trade per person: $288
Adult Literacy: 85%
Life Expectancy (F): 67
Life Expectancy (M): 63

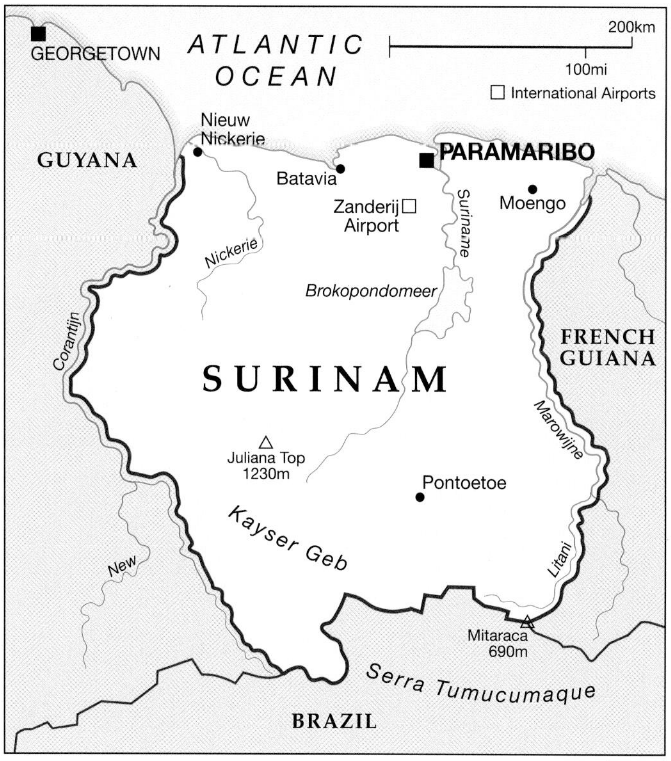

SURINAM

Facts and Figures

Capital: Paramaribo 201,000 (1993)
Area (sq km): 163,820
Population: 404,310 (1991)
Language: Dutch, Spanish, English
Religion: Hindu 24% Roman Catholic 20% Muslim 17% Protestant 16% Other 7%
Currency: Surinam Guilder = 100 cents
Annual Income per person: $3,610
Annual Trade per person: $2,100
Adult Literacy: 95%
Life Expectancy (F): 73
Life Expectancy (M): 68

URUGUAY

Facts and Figures

Capital: Montevideo 1,383,700 (1992)
Area (sq km): 176,215
Population: 3,120,000 (1992)
Language: Spanish
Religion: Roman Catholic 56%
Currency: Uruguayan peso = 100 centésimos
Annual Income per person: $2,860
Annual Trade per person: $1,032
Adult Literacy: 96%
Life Expectancy (F): 76
Life Expectancy (M): 69

VENEZUELA

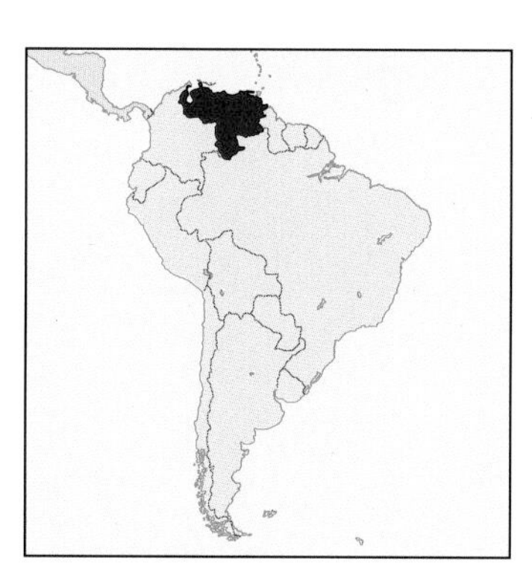

Facts and Figures

Capital: Caracas 1,044,900 (1981)
Area (sq km): 912,050
Population: 20,410,000 (1993)
Language: Spanish
Religion: Roman Catholic 86%
Currency: Bolívar = 100 céntimos
Annual Income per person: $2,610
Annual Trade per person: $1,288
Adult Literacy: 88%
Life Expectancy (F): 74
Life Expectancy (M): 67

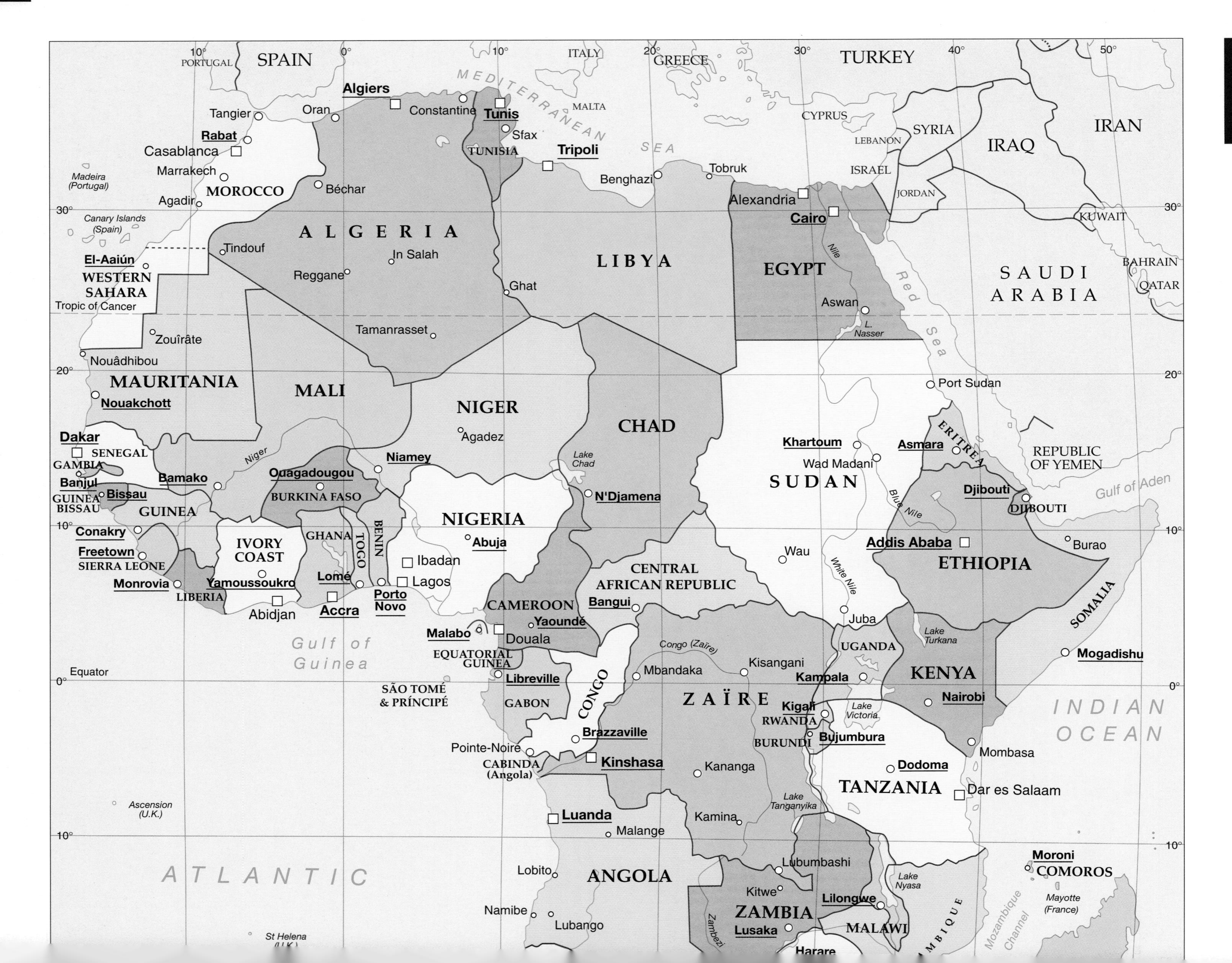
PORTUGAL
SPAIN
ITALY
GREECE
TURKEY
CYPRUS
SYRIA
LEBANON
ISRAEL
JORDAN
IRAQ
IRAN
KUWAIT
BAHRAIN
QATAR
SAUDI ARABIA
REPUBLIC OF YEMEN
MALTA
MEDITERRANEAN SEA
Red Sea
Gulf of Aden
INDIAN OCEAN
ATLANTIC
Gulf of Guinea
MOROCCO
Rabat
Tangier
Casablanca
Marrakech
Agadir
Madeira (Portugal)
Canary Islands (Spain)
ALGERIA
Algiers
Oran
Constantine
Béchar
Tindouf
Reggane
In Salah
Tamanrasset
TUNISIA
Tunis
Sfax
LIBYA
Tripoli
Benghazi
Tobruk
Ghat
EGYPT
Cairo
Alexandria
Aswan
L. Nasser
Nile
WESTERN SAHARA
El-Aaiún
Tropic of Cancer
MAURITANIA
Nouakchott
Nouâdhibou
Zouîrâte
MALI
Bamako
Niger
NIGER
Niamey
Agadez
CHAD
N'Djamena
Lake Chad
SUDAN
Khartoum
Wad Madani
Port Sudan
Wau
Juba
Blue Nile
White Nile
ERITREA
Asmara
DJIBOUTI
Djibouti
ETHIOPIA
Addis Ababa
SOMALIA
Mogadishu
Burao
SENEGAL
Dakar
GAMBIA
Banjul
GUINEA BISSAU
Bissau
GUINEA
Conakry
SIERRA LEONE
Freetown
LIBERIA
Monrovia
IVORY COAST
Yamoussoukro
Abidjan
BURKINA FASO
Ouagadougou
GHANA
Accra
TOGO
Lomé
BENIN
Porto Novo
NIGERIA
Abuja
Ibadan
Lagos
CAMEROON
Yaoundé
Douala
EQUATORIAL GUINEA
Malabo
SÃO TOMÉ & PRÍNCIPE
GABON
Libreville
CENTRAL AFRICAN REPUBLIC
Bangui
CONGO
Brazzaville
Pointe-Noire
CABINDA (Angola)
ZAÏRE
Kinshasa
Congo (Zaïre)
Mbandaka
Kisangani
Kananga
Kamina
Lubumbashi
UGANDA
Kampala
KENYA
Nairobi
Mombasa
Lake Turkana
Lake Victoria
RWANDA
Kigali
BURUNDI
Bujumbura
TANZANIA
Dodoma
Dar es Salaam
Lake Tanganyika
Lake Nyasa
ANGOLA
Luanda
Malange
Lobito
Lubango
Namibe
ZAMBIA
Lusaka
Kitwe
Zambezi
MALAWI
Lilongwe
Harare
COMOROS
Moroni
Mayotte (France)
Mozambique Channel
Ascension (U.K.)
Equator
0°
10°
20°
30°
40°
50°

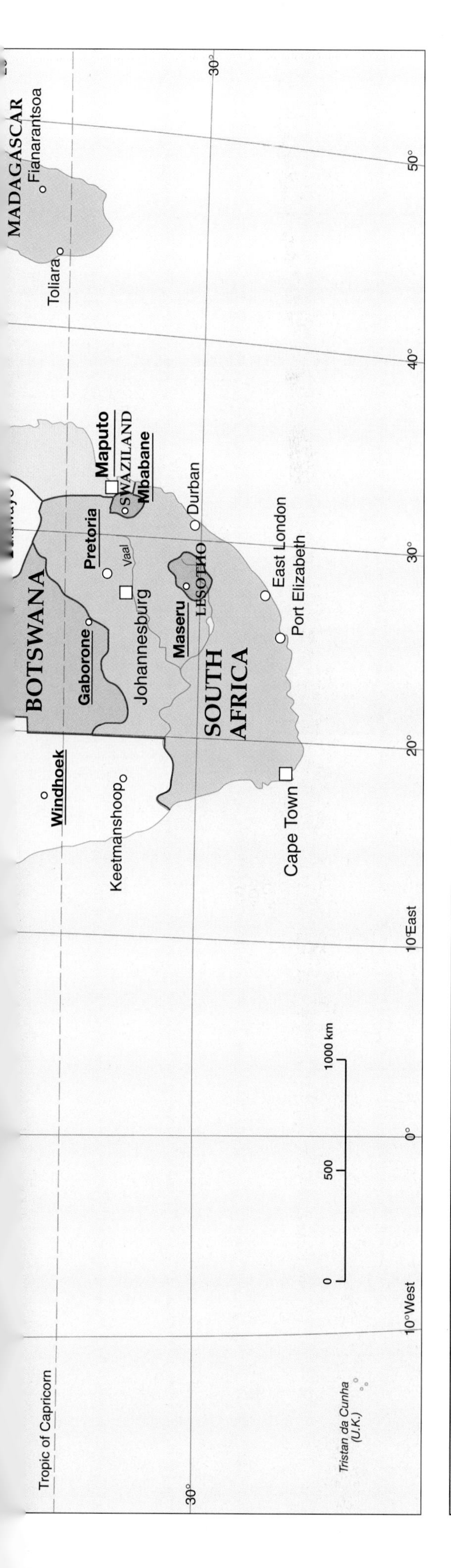
MADAGASCAR
Fianarantsoa
Toliara
Maputo
SWAZILAND
Mbabane
Durban
Pretoria
Vaal
East London
Port Elizabeth
BOTSWANA
Gaborone
Johannesburg
Maseru
LESOTHO
SOUTH AFRICA
Windhoek
Keetmanshoop
Cape Town
Tristan da Cunha (U.K.)
Tropic of Capricorn
50°
40°
30°
20°
10°East
0°
10°West
30°
0
500
1000 km

COUNTRY	PAGE
Algeria	50
Angola	50
Azores	50
Benin	51
Botswana	51
Burkina Faso	51
Burundi	52
Cameroon	52
Canary Islands	52
Cape Verde	53
Central African Republic	53
Chad	53
Comoros	54
Congo	54
Cote d'Ivoire	54
Djibouti	54
Egypt	55
Eritrea	55
Ethiopia	55
Equatorial Guinea	55
Gabon	56
Gambia	56
Ghana	56
Guinea	57
Guinea Bissau	57
Kenya	57
Lesotho	58
Liberia	58
Libya	58
Madagascar	59
Madeira	59
Malawi	59
Mali	60
Mauritania	60
Mauritius	60
Morocco	61
Mozambique	61
Namibia	61
Niger	62
Nigeria	62
Rwanda	62
São Tomé & Príncipe	63
Senegal	63
Seychelles	63
Sierra Leone	63
Somalia	64
South Africa	64
Sudan	64
Swaziland	65
Tanzania	65
Togo	65
Tunisia	66
Uganda	66
Zaire	66
Zambia	67
Zimbabwe	67

ALGERIA

Facts and Figures

Capital: Algiers 1,507,000 (1987)
Area (sq km): 2,381,741
Population: 26,600,000 (1993)
Language: Arabic, French, Berber
Religion: Sunni Muslim 98%
Currency: Algerian Dinar = 100 centimes
Annual Income per person: $2,020
Annual Trade per person: $639
Adult Literacy: 57%
Life Expectancy (F): 67
Life Expectancy (M): 65

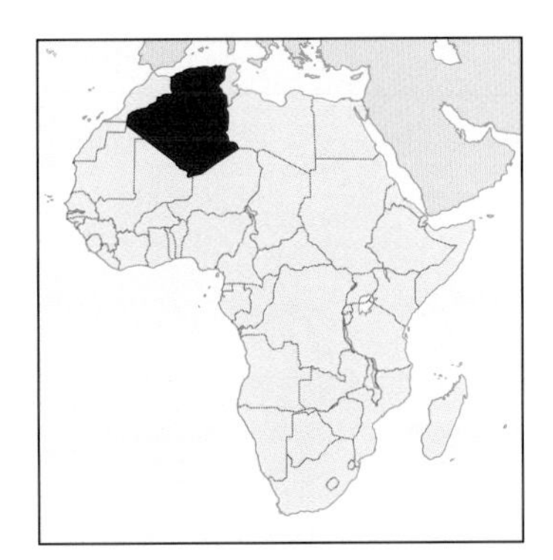

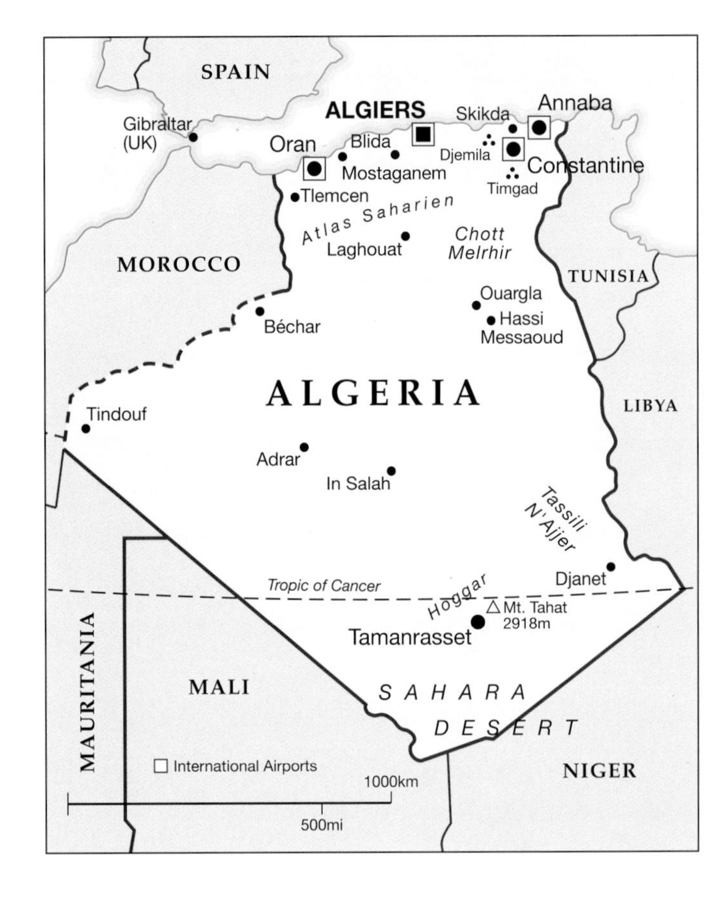

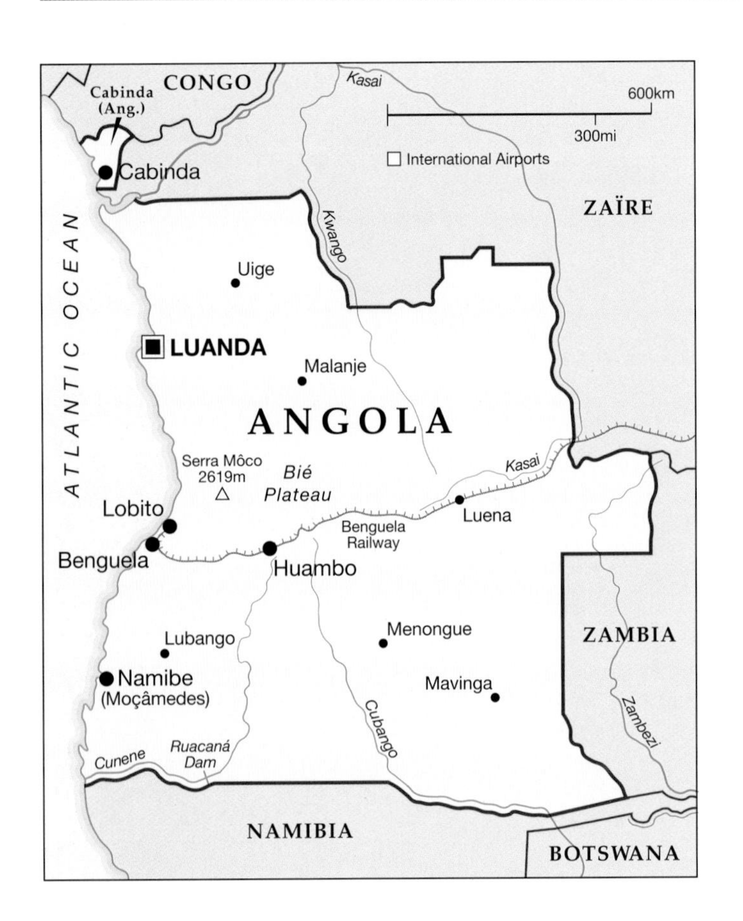

ANGOLA

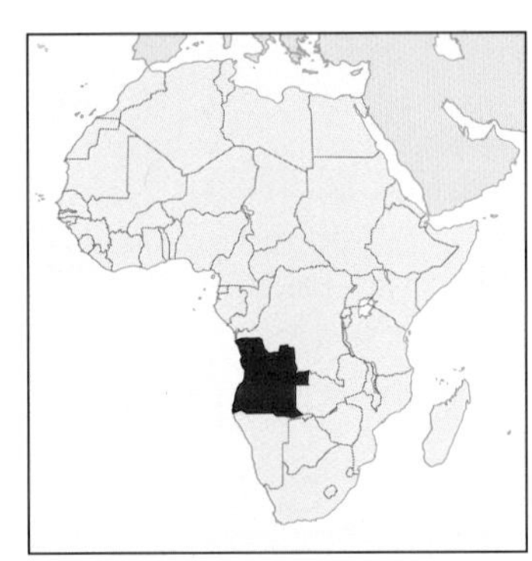

Facts and Figures

Capital: Luanda 480,600 (1988)
Area (sq km): 1,246,700
Population: 10,770,000 (1993)
Language: Portuguese
Religion: Roman Catholic 75%
Protestant 13% Traditional 16%
Currency: Kwanza = 100 lwei
Annual Income per person: $620
Annual Trade per person: $281
Adult Literacy: 42%
Life Expectancy (F): 48
Life Expectancy (M): 45

AZORES

Facts and Figures

Capital: Ponta Delgada 21,091 (1991)
Area (sq km): 2,247
Population: 260,000 (1991)
Language: Portuguese
Religion: Roman Catholic
Currency: Escudo
Annual Income per person: N/A
Annual Trade per person: N/A
Adult Literacy: N/A
Life Expectancy (F): N/A
Life Expectancy (M): N/A

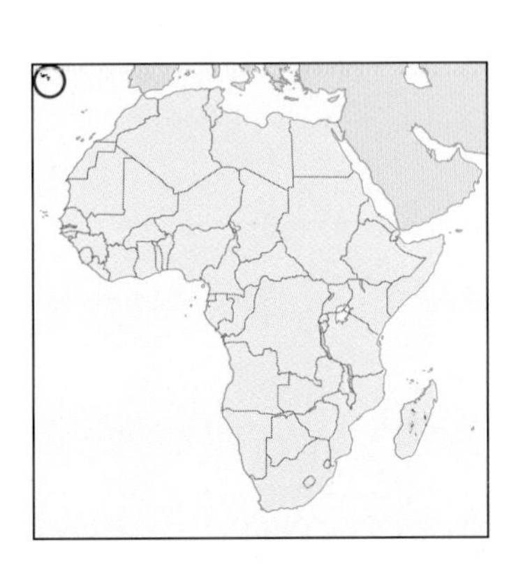

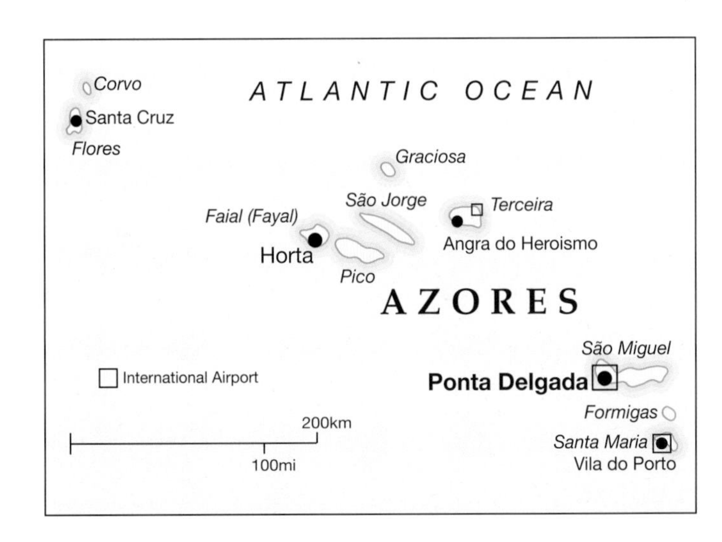

NIGER
BURKINA FASO
Oti
Pendjari Nat. Park
Atakora Mts.
Mekrou
Alibori
Kandi
Niger
BENIN
Natitingou
Djougou
Parakou
Kainji Reservoir
TOGO
Ouémé
NIGERIA
GHANA
Savé
Savalou
□ International Airports
200km
100mi
Mono
Abomey
PORTO NOVO
Ouidah
Cotonou
Lagos
Bight of Benin

BENIN.

Facts and Figures

Capital: Porto Novo 208,260 (1982)
Area (sq km): 112,620
Population: 5,010,000 (1993)
Language: French, Fon
Religion: Traditional 60% Roman Catholic 18% Sunni Muslim 15%
Currency: CFA Franc = 100 centimes
Annual Income per person: $380
Annual Trade per person: $156
Adult Literacy: 23%
Life Expectancy (F): 50
Life Expectancy (M): 46

BOTSWANA

Facts and Figures

Capital: Gaborone 138,470 (1991)
Area (sq km): 581,730
Population: 1,330,000 (1991)
Language: English, Setswana
Religion: Traditional 50% Christian 30%
Currency: Pula = 100 thebe
Annual Income per person: $2,590
Annual Trade per person: $2,500
Adult Literacy: 74%
Life Expectancy (F): 64
Life Expectancy (M): 58

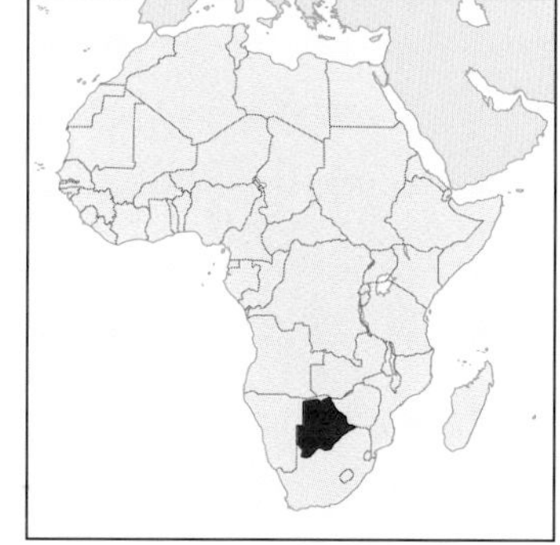

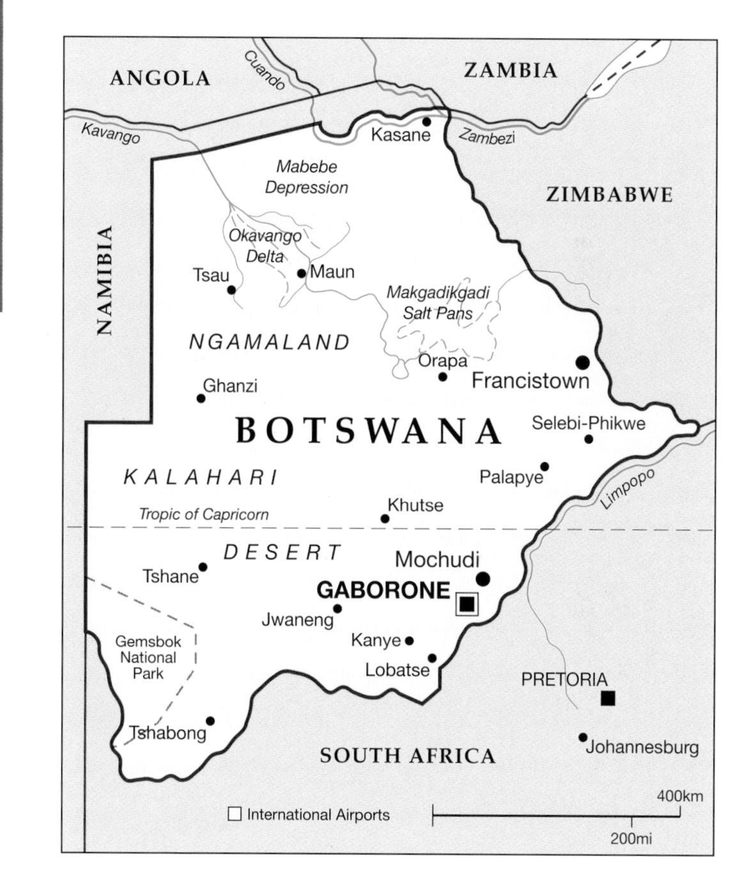

400km
200mi
□ International Airports
NIGER
Niger
Ouahigouya
MALI
BURKINA FASO
OUAGADOUGOU
Black Volta
Koudougou
Fada N'Gourma
Red Volta
Tenkodogo
Bobo Dioulasso
BENIN
GHANA
TOGO
White Volta
CÔTE D'IVOIRE

BURKINA FASO

Facts and Figures

Capital: Ouagadougou 442,230 (1985)
Area (sq km): 274,400
Population: 9,190,000 (1991)
Language: French
Religion: Sunni Muslim 52% Christian 21%
Currency: CFA Franc = 100 centimes
Annual Income per person: $350
Annual Trade per person: $48
Adult Literacy: 18%
Life Expectancy (F): 51
Life Expectancy (M): 48

BURUNDI

Facts and Figures

Capital: Bujumbura 241,000 (1989)
Area (sq km): 27,830
Population: 5,600,000 (1992)
Language: French, Kirundi
Religion: Roman Catholic 62%
Traditional 30%
Currency: Burundi Franc = 100 centimes
Annual Income per person: $210
Annual Trade per person: $60
Adult Literacy: 50%
Life Expectancy (F): 51
Life Expectancy (M): 48

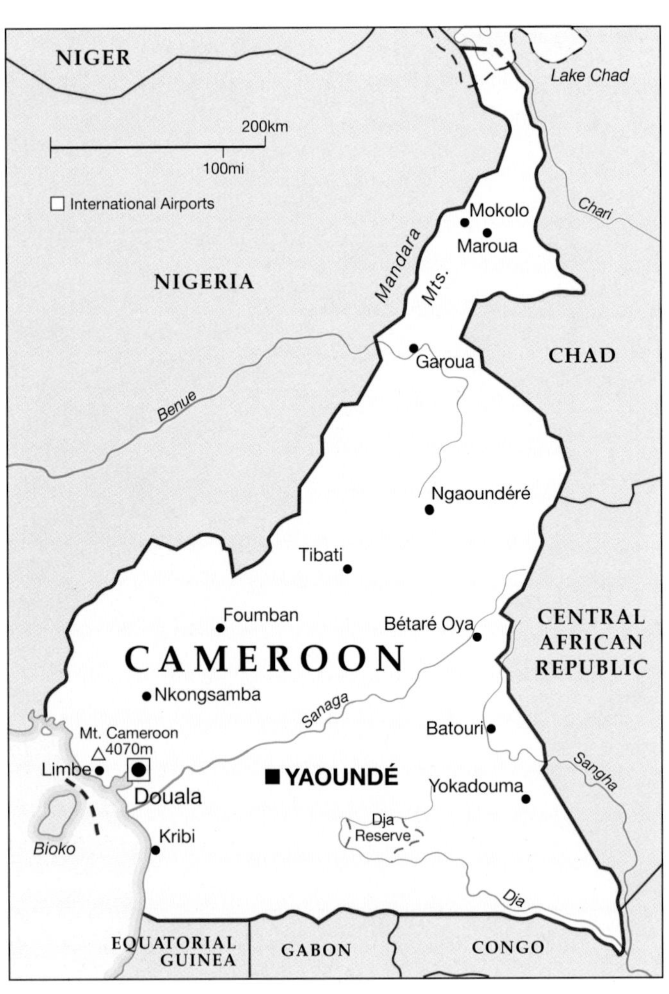

CAMEROON

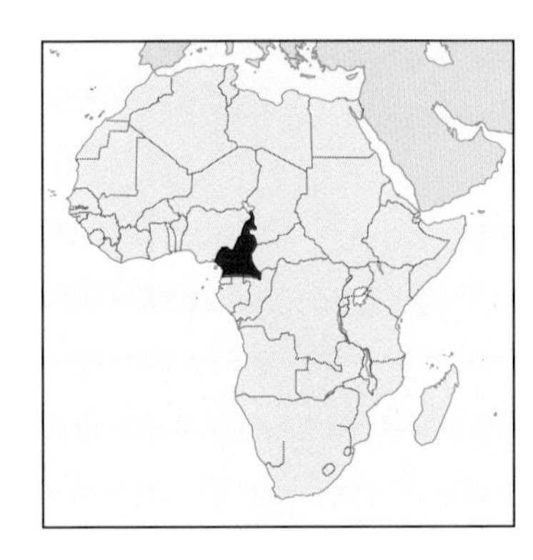

Facts and Figures

Capital: Yaoundé 750,000 (1991)
Area (sq km): 475,442
Population: 12,240,000 (1991)
Language: French, English
Religion: Roman Catholic 28%
Animist 25% Sunni Muslim 22%
Protestant 18%
Currency: CFA Franc = 100 centimes
Annual Income per person: $940
Annual Trade per person: $310
Adult Literacy: 54%
Life Expectancy (F): 57
Life Expectancy (M): 54

CANARY ISLANDS

Facts and Figures

Facts and Figures for CANARY ISLANDS are incorporated with SPAIN

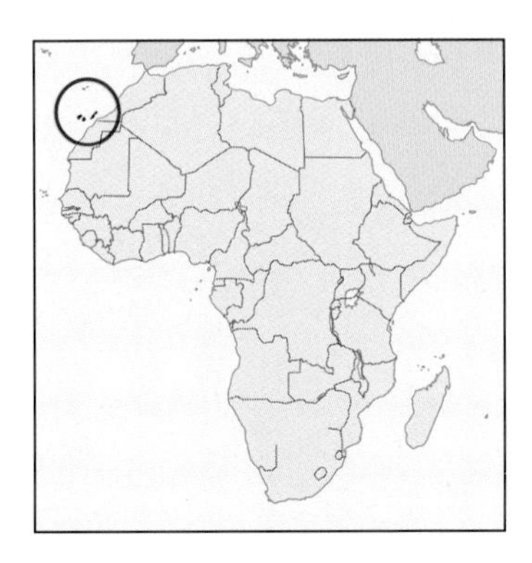

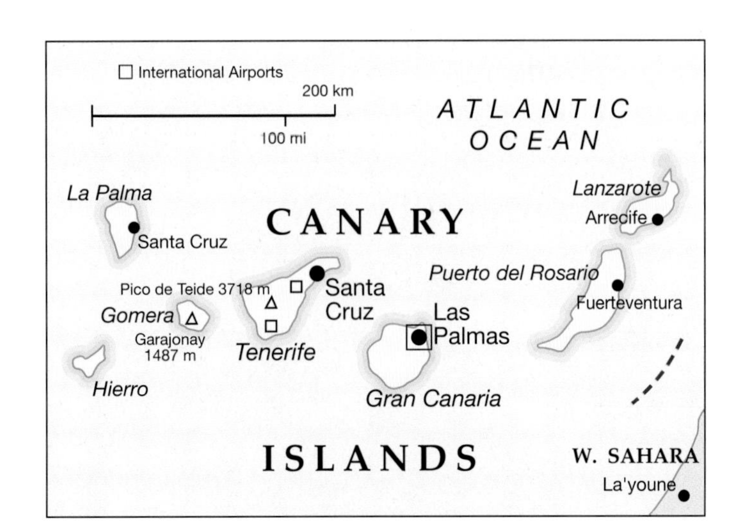

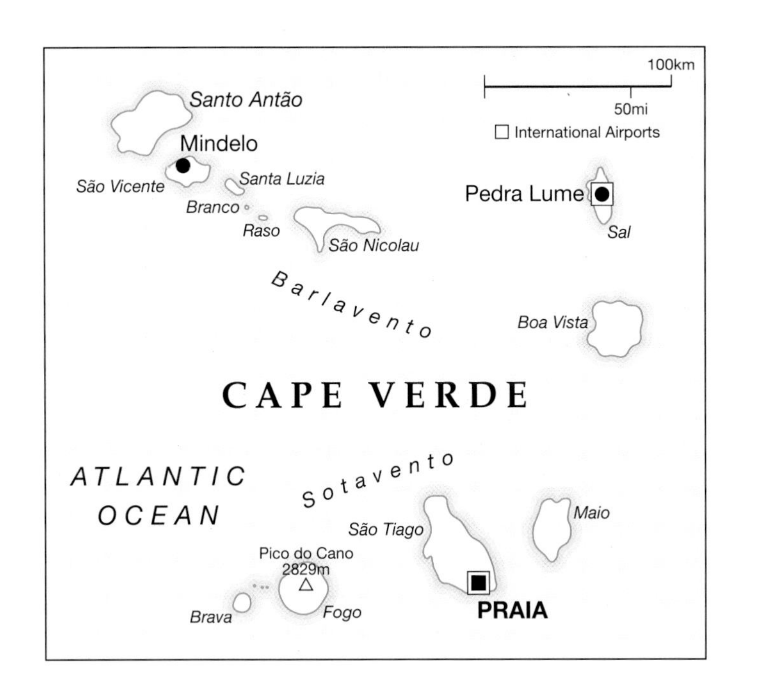

CAPE VERDE

Facts and Figures

Capital: Praia 61,700 (1990)
Area (sq km): 4,030
Population: 350,000 (1993)
Language: Portuguese, Crioulo
Religion: Roman Catholic 93%
Protestant 7%
Currency: Escudo = 100 centavos
Annual Income per person: $750
Annual Trade per person: $331
Adult Literacy: 53 %
Life Expectancy (F): 69
Life Expectancy (M): 67

CENTRAL AFRICAN REPUBLIC

Facts and Figures

Capital: Bangui 451,700 (1988)
Area (sq km): 622,440
Population: 3,130,000 (1991)
Language: French, Sango
Religion: Protestant 45% RC 30%
Traditional
Currency: CFA Franc = 100 centimes
Annual Income per person: $390
Annual Trade per person: $70
Adult Literacy: 38%
Life Expectancy (F): 53
Life Expectancy (M): 48

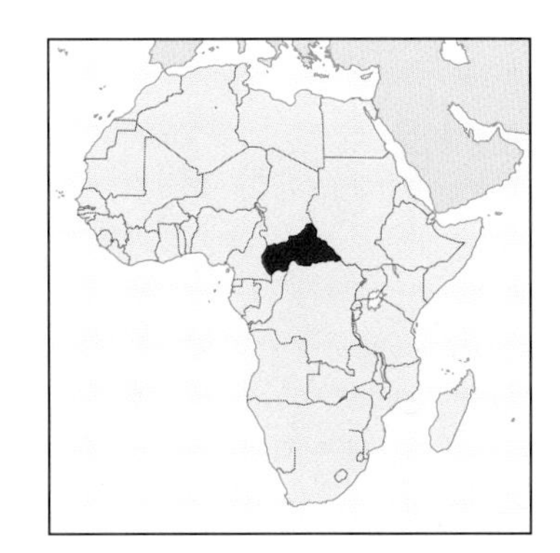

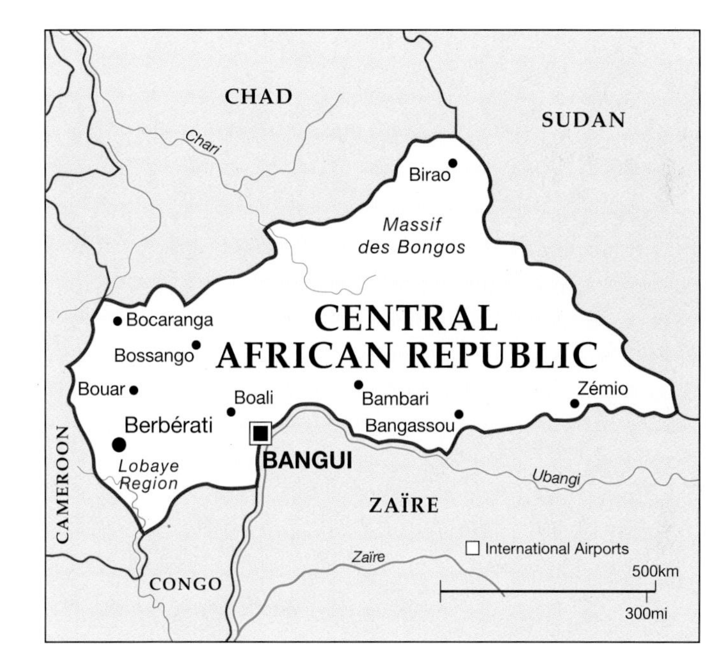

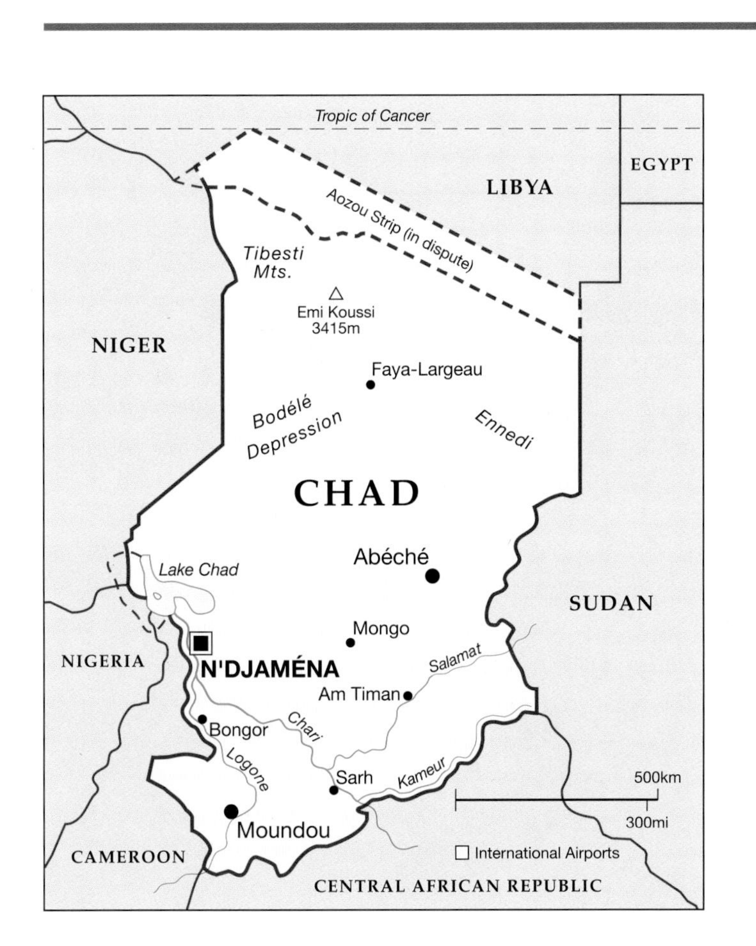

CHAD

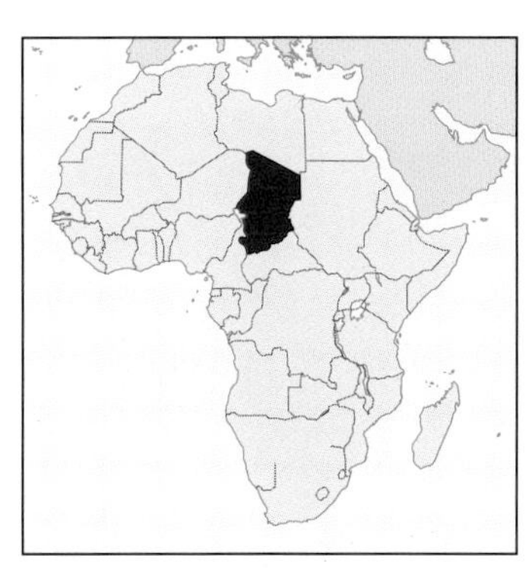

Facts and Figures

Capital: N'Djaména 529,560 (1993)
Area (sq km): 1,284,000
Population: 6,290,000 (1993)
Language: French, Arabic
Religion: Muslim 41% Traditional 35%
Christian 23%
Currency: CFA Franc = 100 centimes
Annual Income per person: $220
Annual Trade per person: $104
Adult Literacy: 30%
Life Expectancy (F): 49
Life Expectancy (M): 46

COMOROS

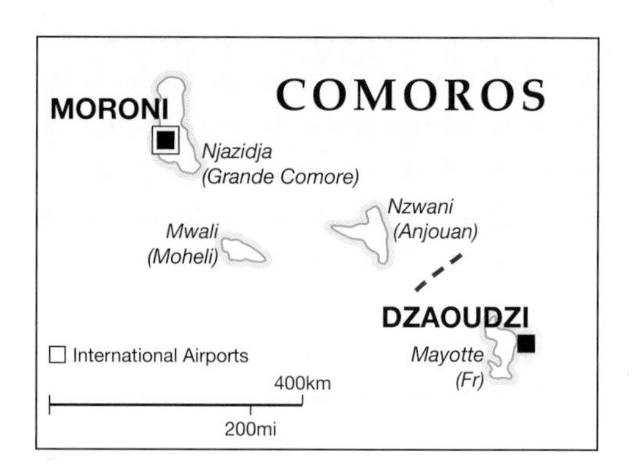

Facts and Figures

Capital: Moroni 22,000 (1992)
Area (sq km): 2,230
Population: 510,000 (1992)
Language: Swahili, French, Arabic
Religion: Sunni Muslim 93%
Currency: Comorian Franc
Annual Income per person: $500
Annual Trade per person: $120
Adult Literacy: 61%
Life Expectancy (F): 57
Life Expectancy (M): 56

CONGO

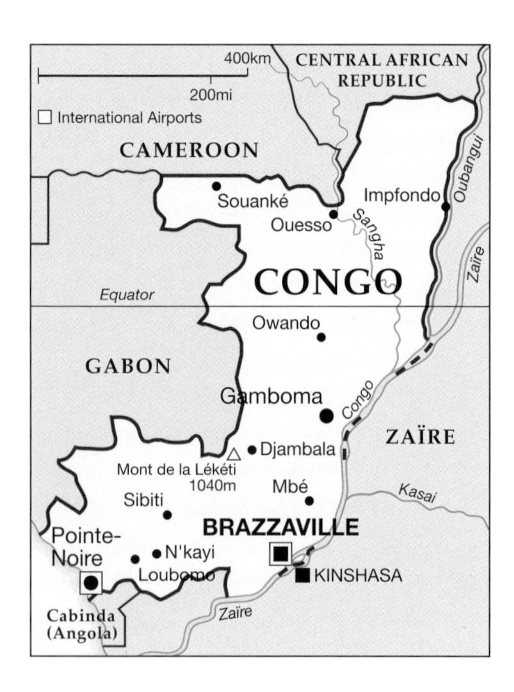

Facts and Figures

Capital: Brazzaville 937,600 (1992)
Area (sq km): 341,800
Population: 2,690,000 (1992)
Language: French
Religion: Roman Catholic 46% Protestant 19% Traditional
Currency: CFA Franc = 100 centimes
Annual Income per person: $1,120
Annual Trade per person: $694
Adult Literacy: 57%
Life Expectancy (F): 57
Life Expectancy (M): 52

COTE D'IVOIRE

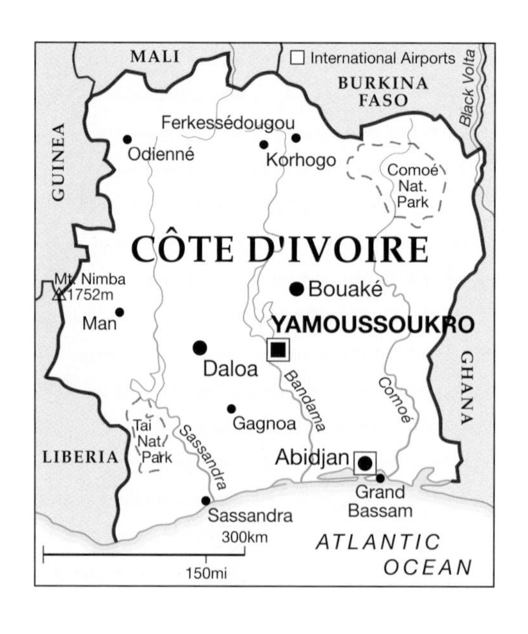

Facts and Figures

Capital: Yamoussoukro 120,000 (1986)
Area (sq km): 322,460
Population: 13,100,000 (1991)
Language: French
Religion: Muslim 20% RC 20% Traditional 44%
Currency: CFA Franc = 100 centimes
Annual Income per person: $690
Annual Trade per person: $443
Adult Literacy: 54%
Life Expectancy (F): 56
Life Expectancy (M): 53

DJIBOUTI

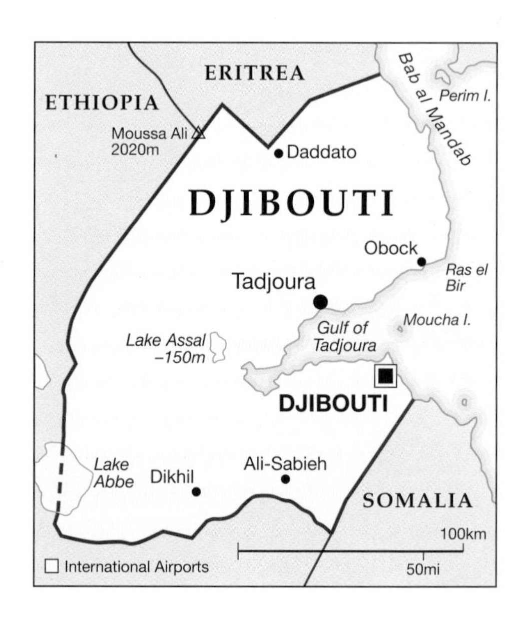

Facts and Figures

Capital: Djibouti 317,000 (1991)
Area (sq km): 23,200
Population: 542,000 (1991)
Language: Arabic, French, Cushitic
Religion: Sunni Muslim 94% Roman Catholic 4%
Currency: Djibouti Franc
Annual Income per person: $1,000
Annual Trade per person: $550
Adult Literacy: 19%
Life Expectancy (F): 51
Life Expectancy (M): 47

EGYPT

MEDITERRANEAN SEA
Port Said
Alexandria
Matrûh
El Alamein
Abu Mina
CAIRO
Suez
Suez Canal
JORDAN
ISRAEL
Giza
Siwa
Qattara Depression –133m
Giza Pyramids & Sphinx
Sinai
SAUDI ARABIA
El Minya
Nile
LIBYA
EGYPT
Hurghada
RED SEA
Valley of the Kings
Thebes
Luxor
International Airports
El Kharga
400km
200mi
The Great Oasis
Aswan
dam
Philae
Tropic of Cancer
Lake Nasser
Admin. boundary
SAHARA DESERT
Abu Simbel
Political boundary
SUDAN

Facts and Figures

Capital: Cairo 6,663,000 (1991)
Area (sq km): 1,001,450
Population: 56,430,000 (1993)
Language: Arabic, French, English
Religion: Sunni Muslim 92% Christian
Currency: Egyptian Pound = 100 piastres
Annual Income per person: $620
Annual Trade per person: $211
Adult Literacy: 48%
Life Expectancy (F): 63
Life Expectancy (M): 60

EQUATORIAL GUINEA

Facts and Figures

Capital: Malabo 370,000 (1988)
Area (sq km): 28,051
Population: 417,000 (1990)
Language: Spanish, Fang
Religion: Roman Catholic 89%
Currency: CFA Franc = 100 centimes
Annual Income per person: $330
Annual Trade per person: $250
Adult Literacy: 50%
Life Expectancy (F): 50
Life Expectancy (M): 46

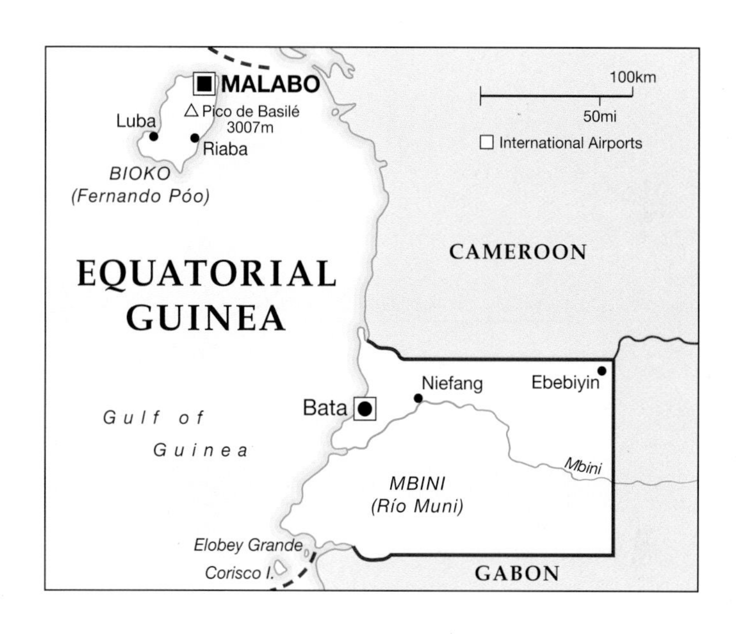

ERITREA & ETHIOPIA

Facts and Figures

ETHIOPIA

Capital: Addis Ababa 1,700,000 (1990)
Area (sq km): 1,157,603
Population: 51,980,000 (1993)
Language: Amharic, Galla, Tigre
Religion: Muslim 45%, Christian 40%, Traditional 12%
Currency: Ethiopean birr = 100 cents
Annual Income per person: $120
Annual Trade per person: $12
Adult Literacy: 66%
Life Expectancy (F): 49
Life Expectancy (M): 45

Facts and Figures

ERITREA

Capital: Asmara 367,300 (1991)
Area (sq km): 93,679
Population: 3,500,00 (1991)
Language: Tigrinya
Religion: Muslim 50% Coptic Christian 50%
Currency: Ethiopean birr = 100 cents
Annual Income per person: N/A
Annual Trade per person: N/A
Adult Literacy: N/A
Life Expectancy (F): N/A
Life Expectancy (M): N/A

GABON

Facts and Figures

Capital: Libreville 350,000 (1983)
Area (sq km): 267,670
Population: 1,010,000 (1993)
Language: French, Bantu
Religion: Roman Catholic 75%, Traditional
Currency: CFA Franc = 100 centimes
Annual Income per person: $3,780
Annual Trade per person: $2,094
Adult Literacy: 61%
Life Expectancy (F): 55
Life Expectancy (M): 52

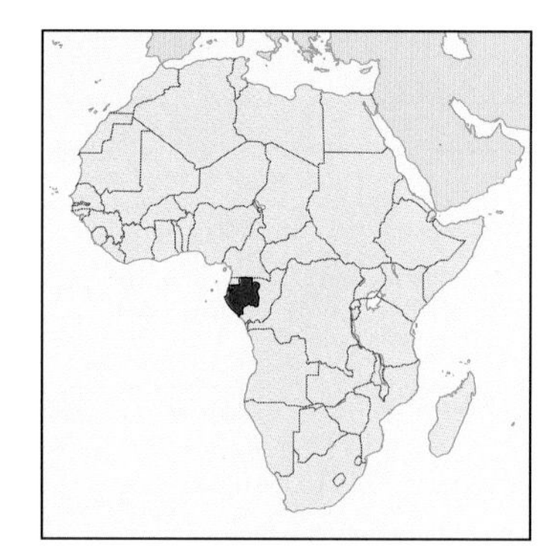

GAMBIA

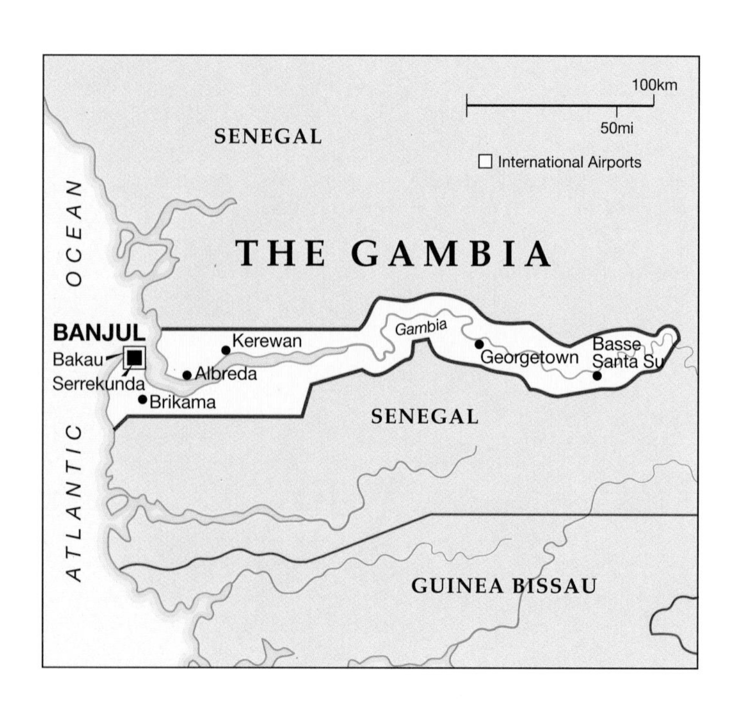

Facts and Figures

Capital: Banjul 52,000 (1992)
Area (sq km): 11,300
Population: 875,000 (1990)
Language: English, Madinka
Religion: Sunni Muslim 95% Christian 4%
Currency: Dalasi = 100 butut
Annual Income per person: $360
Annual Trade per person: $298
Adult Literacy: 27%
Life Expectancy (F): 47
Life Expectancy (M): 43

GHANA

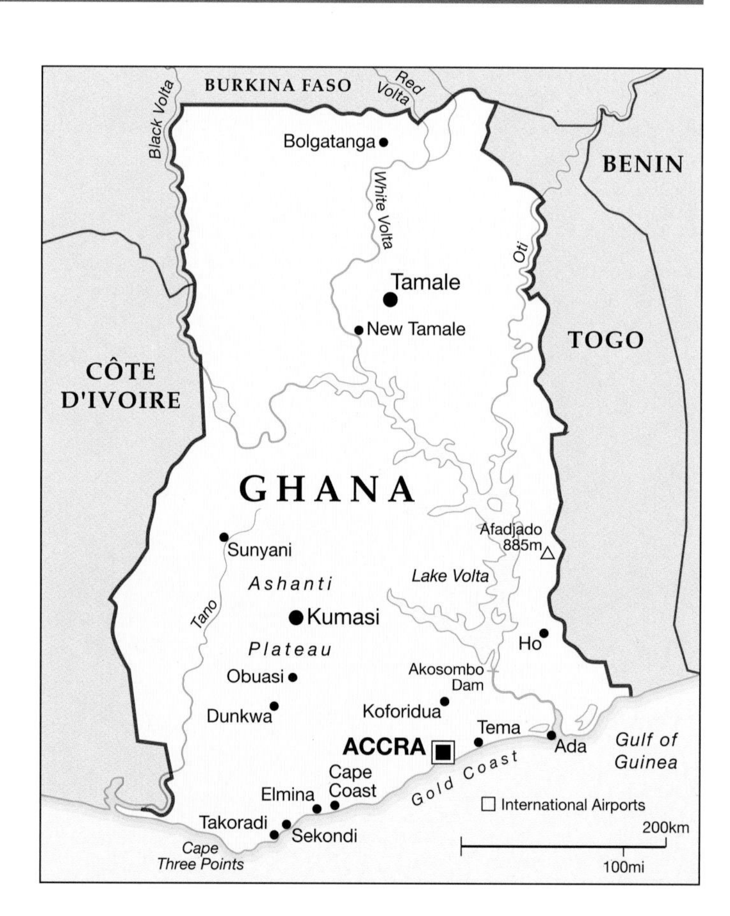

Facts and Figures

Capital: Accra 1,050,000 (1992)
Area (sq km): 238,540
Population: 16,700,000 (1991)
Language: English, Akan, Mossi, Ewe, Ga-Adangme
Religion: Protestant 28% Traditional 21% RC 19% Muslim 16%
Currency: Cedi = 100 pesewas
Annual Income per person: $400
Annual Trade per person: $158
Adult Literacy: 60%
Life Expectancy (F): 58
Life Expectancy (M): 54

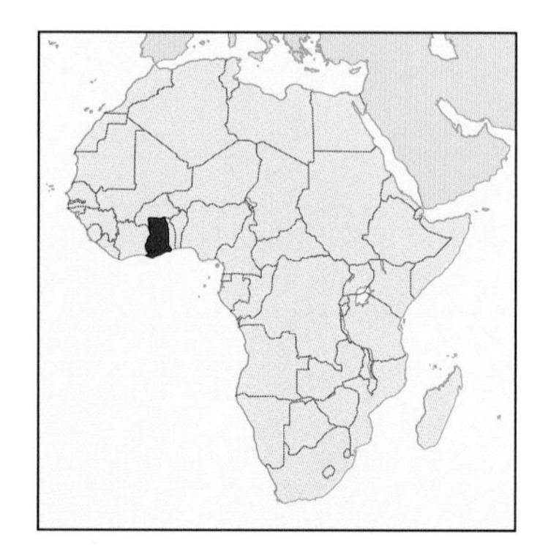

GUINEA

Facts and Figures

Capital: Conakry 860,000 (1992)
Area (sq km): 245,860
Population: 7,300,000 (1990)
Language: French, Susu, Fulani, Malinké
Religion: Muslim 69% Traditional 30%
Currency: Guinean Franc = 100 cauris
Annual Income per person: $450
Annual Trade per person: $160
Adult Literacy: 24%
Life Expectancy (F): 45
Life Expectancy (M): 44

GUINEA BISSAU

Facts and Figures

Capital: Bissau 135,000 (1992)
Area (sq km): 36,125
Population: 980,000 (1991)
Language: Portuguese, Crioulo
Religion: Traditional 65% Muslim 30%
Currency: Peso = 100 centavos
Annual Income per person: $190
Annual Trade per person: $85
Adult Literacy: 36%
Life Expectancy (F): 45
Life Expectancy (M): 42

KENYA

Facts and Figures

Capital: Nairobi 1,500,000 (1992)
Area (sq km): 582,650
Population: 27,900,000 (1991)
Language: Swahili, Kikuyu, English
Religion: Roman Catholic 23%
Other Christian 23%
Protestant 15% Traditional 15%
Currency: Kenya shilling = 100 cents
Annual Income per person: $340
Annual Trade per person: $113
Adult Literacy: 69%
Life Expectancy (F): 63
Life Expectancy (M): 59

LESOTHO

Facts and Figures

Capital: Maseru 126,000 (1992)
Area (sq km): 30,355
Population: 1,830,000 (1991)
Language: Sesotho, English
Religion: Protestant 49%
Roman Catholic 44%
Currency: Loti = 100 lisente
Annual Income per person: $580
Annual Trade per person: $350
Adult Literacy: 60%
Life Expectancy (F): 63
Life Expectancy (M): 54

LIBERIA

Facts and Figures

Capital: Monrovia 540,000 (1992)
Area (sq km): 99,070
Population: 2,830,000 (1992)
Language: English, Mande,
West Atlantic, Kwa
Religion: Sunni Muslim 30%
Protestant 15%
Roman Catholic 5%
Currency: Liberian Dollar
Annual Income per person: $600
Annual Trade per person: $285
Adult Literacy: 40%
Life Expectancy (F): 57
Life Expectancy (M): 54

LIBYA

Facts and Figures

Capital: Tripoli 858,000 (1981)
Area (sq km): 1,759,540
Population: 4,700,000 (1990)
Language: Arabic, Berber
Religion: Sunni Muslim 97%
Currency: Libyan dinar = 1000 millemes
Annual Income per person: $5,500
Annual Trade per person: $3,332
Adult Literacy: 64%
Life Expectancy (F): 65
Life Expectancy (M): 62

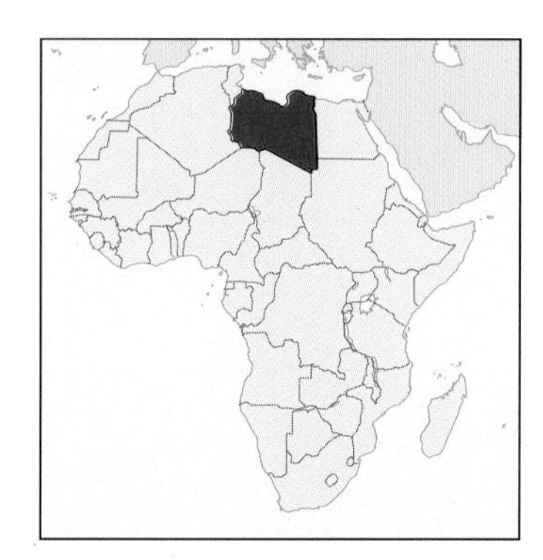

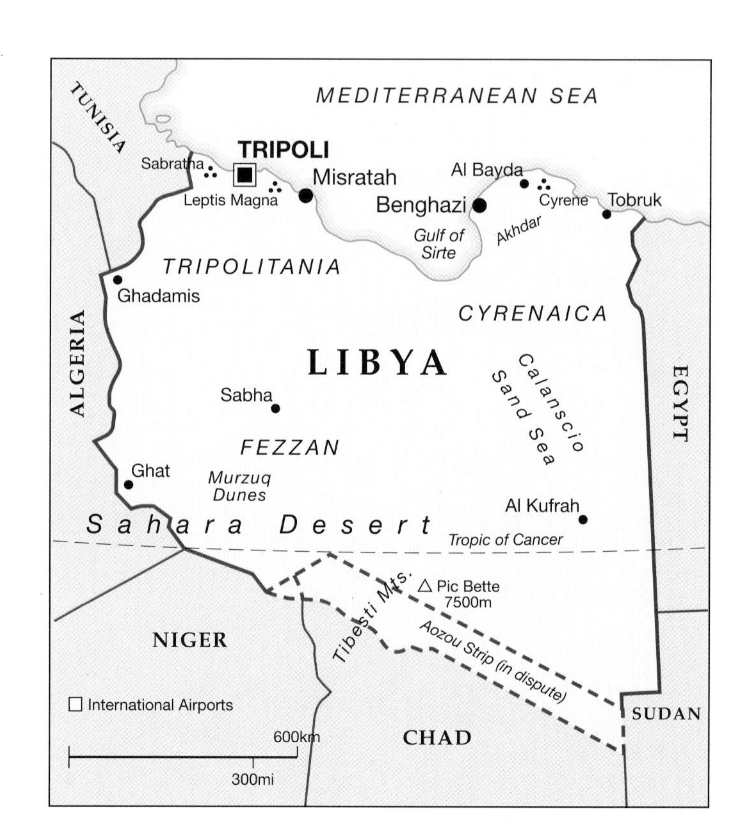

MOZAMBIQUE
Cap d'Ambre
Antseranana
Nosy Bé
Maromokotra 2876m
MADAGASCAR
Mozambique Channel
Mahajanga
Antongil Bay
Nosy Boraha
Betsiboka
Toamasina
Lake Itasy
ANTANANARIVO
Ankaratra 2643m
Vatomandry
Morondava
Bassas da India (Fr)
I. de l'Europa (Fr)
Ambohimanga
Mangoky
Fianarantsoa
INDIAN OCEAN
Isalo Plateau
Toliara
Tropic of Capricorn
International Airports
Toalañaro
400km
200mi
Cap Ste Marie

MADAGASCAR

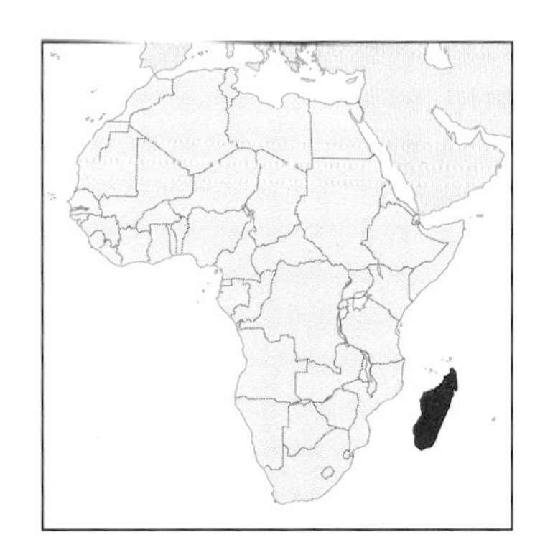

Facts and Figures

Capital: Antananarivo 1,803,390 (1990)
Area (sq km): 587,040
Population: 12,370,000 (1991)
Language: Malagasy, French, English
Religion: Traditional 47% RC 26% Protestant 22%
Currency: Malagasy Franc = 100 centimes
Annual Income per person: $210
Annual Trade per person: $65
Adult Literacy: 80%
Life Expectancy (F): 57
Life Expectancy (M): 54

MADEIRA

Facts and Figures

Facts and Figures for MADEIRA are incorporated with PORTUGAL

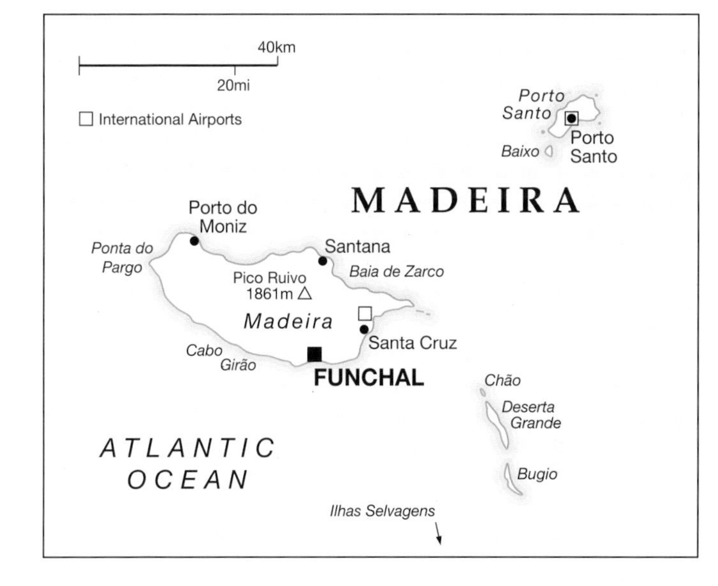

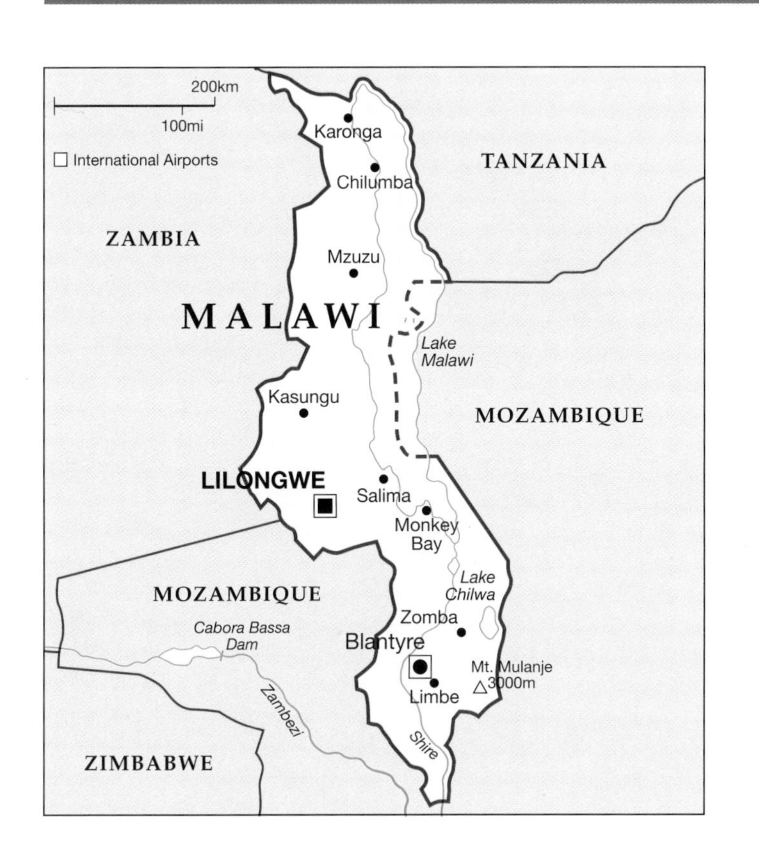

MALAWI

Facts and Figures

Capital: Lilongwe 275,000 (1992)
Area (sq km): 118,480
Population: 9,700,000 (1991)
Language: Chichewe, English
Religion: Protestant 64% RC 17% Muslim 12%
Currency: Kwacha = 100 tambala
Annual Income per person: $230
Annual Trade per person: $138
Adult Literacy: 22%
Life Expectancy (F): 50
Life Expectancy (M): 48

MALI

Facts and Figures

Capital: Bamako 745,000 (1992)
Area (sq km): 1,240,000
Population: 9,362,000 (1990)
Language: French, Bambara
Religion: Sunni Muslim 90% Traditional 9%
Currency: CFA Franc = 100 centimes
Annual Income per person: $280
Annual Trade per person: $97
Adult Literacy: 33%
Life Expectancy (F): 48
Life Expectancy (M): 40

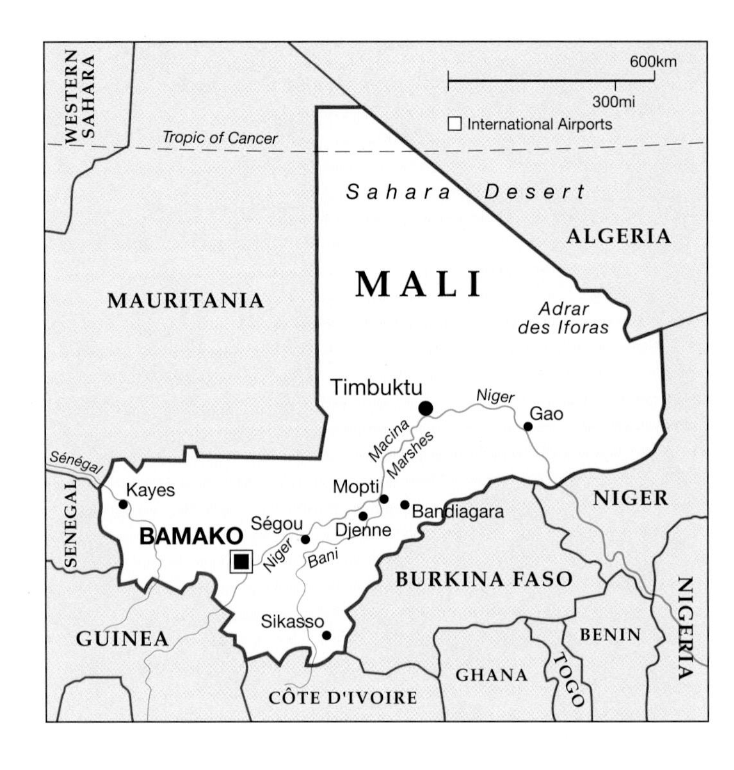

MAURITANIA

Facts and Figures

Capital: Nouakchott 440,600 (1992)
Area (sq km): 1,025,520
Population: 2,110,000 (1992)
Language: Arabic, French, Pulaar, Soninke, Wolof
Religion: Sunni Muslim 99%
Currency: Ouguiya = 5 khoumi
Annual Income per person: $510
Annual Trade per person: $335
Adult Literacy: 34%
Life Expectancy (F): 50
Life Expectancy (M): 46

MAURITIUS

Facts and Figures

Capital: Port Louis 143,000 (1992)
Area (sq km): 2,040
Population: 1,092,400 (1992)
Language: English, Creole, Hindi, French
Religion: Hindu 49% RC 26% Muslim 16%
Currency: Mauritius Rupee = 100 cents
Annual Income per person: $2,420
Annual Trade per person: $2,586
Adult Literacy: 83%
Life Expectancy (F): 73
Life Expectancy (M): 68

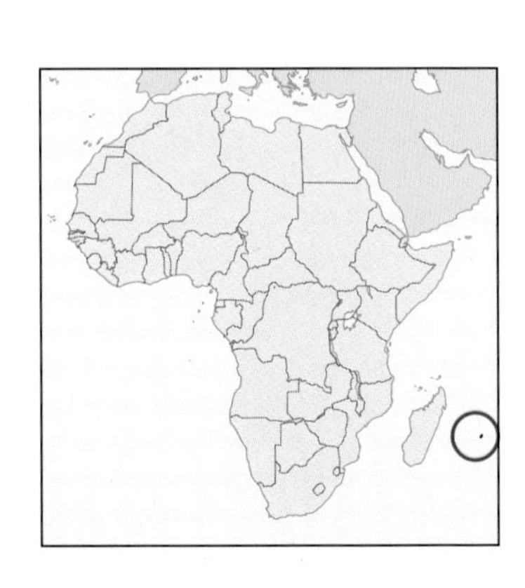

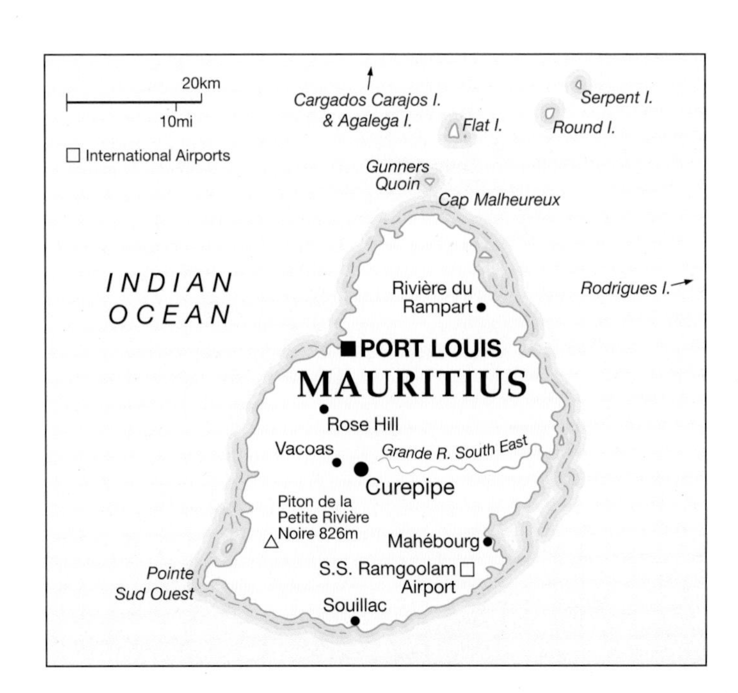

MOROCCO

Facts and Figures

Capital: Rabat 650,000 (1982)
Area (sq km): 458,730
Population: 25,700,00 (1991)
Language: Arabic, Berber
Religion: Sunni Muslim 98%
Currency: Dirham = 100 centimes
Annual Income per person: $1,030
Annual Trade per person: $434
Adult Literacy: 50%
Life Expectancy (F): 65
Life Expectancy (M): 62

MOZAMBIQUE

Facts and Figures

Capital: Maputo 1,098,000 (1991)
Area (sq km): 799,380
Population: 16,110,000 (1991)
Language: Portuguese, Bantu
Religion: Traditional 60% RC 18%
Muslim 13%
Currency: Metical = 100 centavos
Annual Income per person: $70
Annual Trade per person: $66
Adult Literacy: 35%
Life Expectancy (F): 50
Life Expectancy (M): 47

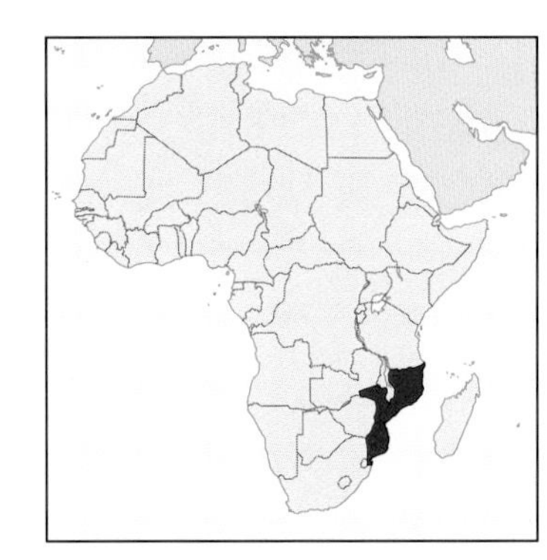

NAMIBIA

Facts and Figures

Capital: Windhoek 159,000 (1991)
Area (sq km): 824,270
Population: 1,510,000 (1992)
Language: English, Afrikaans, German
Religion: Lutheran 51%
Other Christian 49% Animist
Currency: Namibia $ & S. African Rand
Annual Income per person: $1,120
Annual Trade per person: $1,200
Adult Literacy: 38%
Life Expectancy (F): 60
Life Expectancy (M): 58

NIGER

Facts and Figures

Capital: Niamey 450,000 (1992)
Area (sq km): 1,267,000
Population: 8,040,000 (1991)
Language: French, Hausa
Religion: Sunni Muslim 71%
Christian Traditional
Currency: CFA Franc = 100 centimes
Annual Income per person: $300
Annual Trade per person: $87
Adult Literacy: 29%
Life Expectancy (F): 48
Life Expectancy (M): 45

600km
300mi
International Airports
Tropic of Cancer
LIBYA
ALGERIA
Hamada Mangueni
Sahara Desert
Ténéré de Tafassasset
Aïr Massif
NIGER
Ténéré
MALI
Niger
Agadez
Tahoua
Tanout
CHAD
Ayorou
NIAMEY
Maradi
Zinder
Diffa
Lake Chad
Dosso
BURKINA FASO
NIGERIA
BENIN
TOGO
CAMEROON

NIGERIA

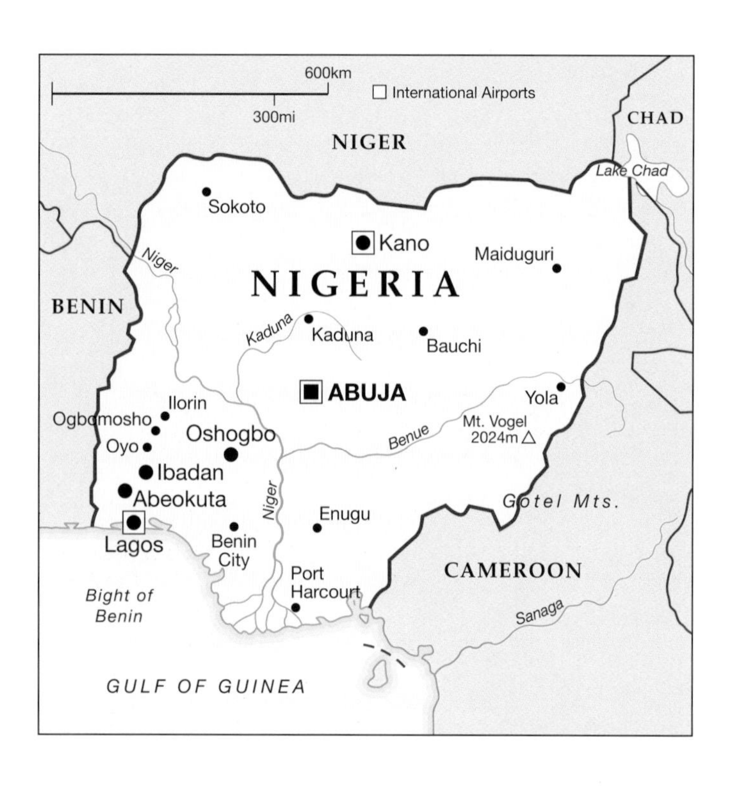

Facts and Figures

Capital: Abuja 305,900 (1992)
Area (sq km): 923,770
Population: 92,800,000 (1991)
Language: English, Hausa, Ibo, Yoruba
Religion: Muslim 48% Protestant 17%
RC 17% Traditional
Currency: Naira = 100 kobo
Annual Income per person: $290
Annual Trade per person: $159
Adult Literacy: 51%
Life Expectancy (F): 54
Life Expectancy (M): 51

RWANDA

Facts and Figures

Capital: Kigali 156,650 (1981)
Area (sq km): 26,340
Population: 7,430,000 (1991)
Language: Kinyarwanda, French,
Kiswahili, Watusi
Religion: Roman Catholic 61%
Protestant 9%
Muslim 9% Traditional
Currency: Rwanda Franc = 100 centimes
Annual Income per person: $260
Annual Trade per person: $70
Adult Literacy: 50%
Life Expectancy (F): 52
Life Expectancy (M): 49

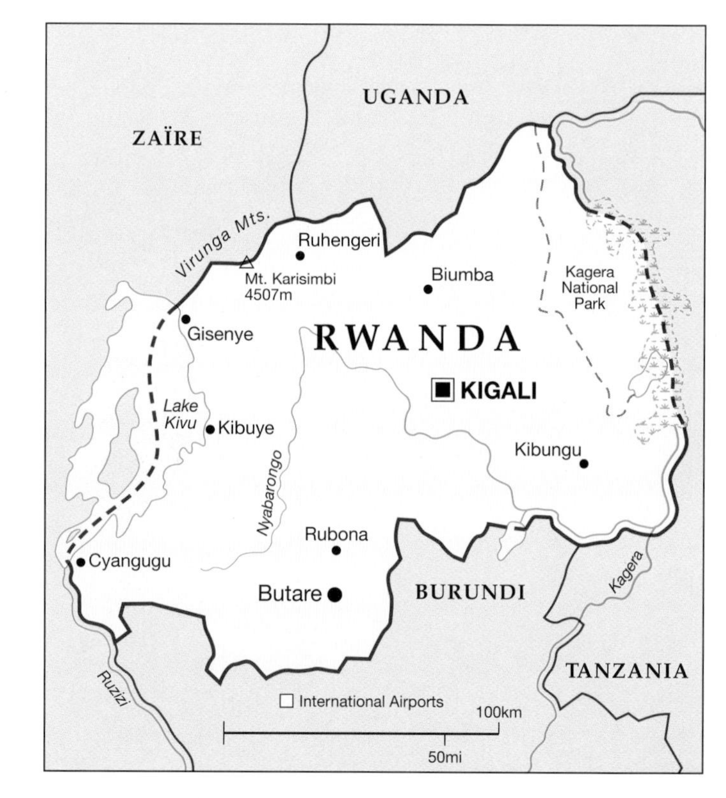

SÃO TOMÉ & PRÍNCIPE

Facts and Figures

Capital: São Tomé 35,000 (1984)
Area (sq km): 1,001
Population: 124,000 (1991)
Language: Portugese, Fang
Religion: Roman Catholic 80%
Currency: Dobra = 100 centimos
Annual Income per person: $350
Annual Trade per person: $200
Adult Literacy: 63%
Life Expectancy (F): 68
Life Expectancy (M): 64

SENEGAL

Facts and Figures

Capital: Dakar 1,730,000 (1992)
Area (sq km): 197,160
Population: 7,970,000 (1993)
Language: French, Wolof
Religion: Sunni Muslim 91%
Christian Animist
Currency: CFA Franc = 100 centimes
Annual Income per person: $720
Annual Trade per person: $235
Adult Literacy: 38%
Life Expectancy (F): 50
Life Expectancy (M): 48

SEYCHELLES

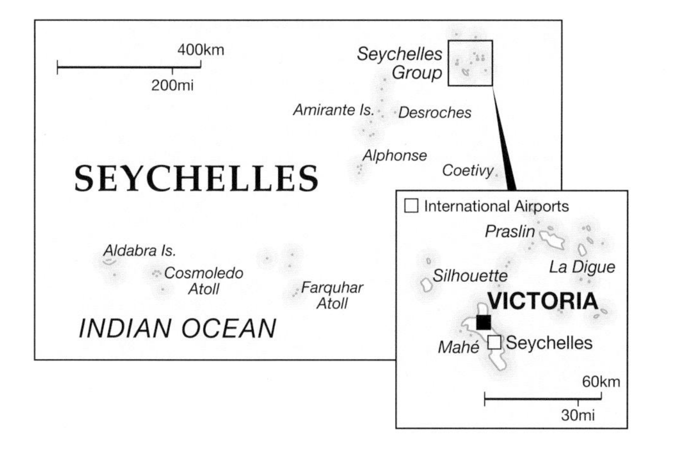

Facts and Figures

Capital: Victoria 24,000 (1992)
Area (sq km): 455
Population: 80,000 (1991)
Language: English, French, Creole
Religion: Roman Catholic 92%
Currency: Seychelles Rupee = 100 cents
Annual Income per person: $5,110
Annual Trade per person: $3,157
Adult Literacy: 89%
Life Expectancy (F): 75
Life Expectancy (M): 65

SIERRA LEONE

Facts and Figures

Capital: Freetown 550,000 (1992)
Area (sq km): 73,330
Population: 4,260,000 (1991)
Language: English, Creole
Religion: Traditional 52% Muslim 39%
Christian
Currency: Leone = 100 cents
Annual Income per person: $210
Annual Trade per person: $72
Adult Literacy: 21%
Life Expectancy (F): 45
Life Expectancy (M): 41

SOMALIA

Facts and Figures

Capital: Mogadishu 485,000 (1992)
Area (sq km): 637,660
Population: 8,000,000 (1990)
Language: Somali, Arabic, English, Italian
Religion: Sunni Muslim 99%
Currency: Somali Shilling = 100 cents
Annual Income per person: $150
Annual Trade per person: $35
Adult Literacy: 25%
Life Expectancy (F): 49
Life Expectancy (M): 45

ERITREA
YEMEN
DJIBOUTI
Gulf of Aden
Caluula
Raas Caseyr
Xaafuun
Mt. Shimbiris 2416m
Berbera
Hargeysa
Burao
Qardho
ETHIOPIA
SOMALIA
Ogaden
Gaalkacyo
Hobyo
Beledweyne
Luuq
Jubba
MOGADISHU
Webi Shabeelle
Marka
INDIAN OCEAN
KENYA
Equator
Kismaayo
International Airports
600km
300mi

SOUTH AFRICA

Facts and Figures

Capital: Pretoria 525,600 (1991)
Area (sq km): 1,126,771
Population: 42,500,000 (1993)
Language: Afrikaans, English, Xhosa, Zulu
Religion: Black Christian 42%
Dutch Reform 23%
Roman Catholic 15%
Currency: Rand = 100 cents
Annual Income per person: $2,600
Annual Trade per person: $964
Adult Literacy: 70%
Life Expectancy (F): 66
Life Expectancy (M): 60

SUDAN

Facts and Figures

Capital: Khartoum 625,000 (1992)
Area (sq km): 2,505,810
Population: 30,830,000 (1993)
Language: Arabic, Nubian, English
Religion: Sunni Muslim 71%
Traditional 16% Christian 8%
Currency: Dinar = 100 piastres
Annual Income per person: $450
Annual Trade per person: $66
Adult Literacy: 27%
Life Expectancy (F): 53
Life Expectancy (M): 51

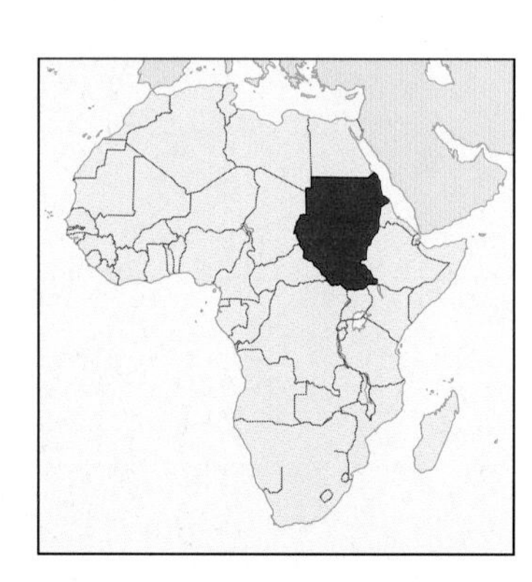

International Airports
100km
50mi
Emblembe 1862m
Pigg's Peak
MAPUTO
Komati
SOUTH AFRICA
Mbuluzi
MBABANE
MOZAMBIQUE
Ezulwini
Lobamba
Siteki
Usutu
Manzini
Maputo
Mankayane
SWAZILAND
Hlatikulu
Ngwavuma
Nhlangano
Lavumisa

SWAZILAND

Facts and Figures

Capital: Mbabane 45,000 (1992)
Area (sq km): 17,360
Population: 835,000 (1989)
Language: Swazi, English
Religion: Christian 60% Traditional 40%
Currency: Lilangeni = 100 cents
Annual Income per person: $1,060
Annual Trade per person: $1,300
Adult Literacy: 72%
Life Expectancy (F): 60
Life Expectancy (M): 56

TANZANIA

Facts and Figures

Capital: Dodoma 225,000 (1992)
Area (sq km): 945,090
Population: 28,200,000 (1991)
Language: Swahili, English
Religion: Christian 40% Muslim 33%
Traditional 23%
Currency: Tanzanian Shilling = 100 cents
Annual Income per person: $100
Annual Trade per person: $56
Adult Literacy: 65%
Life Expectancy (F): 55
Life Expectancy (M): 50

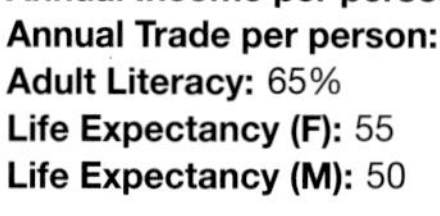

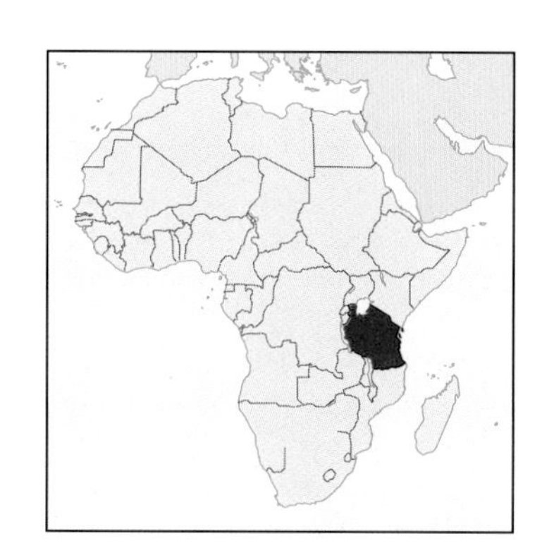

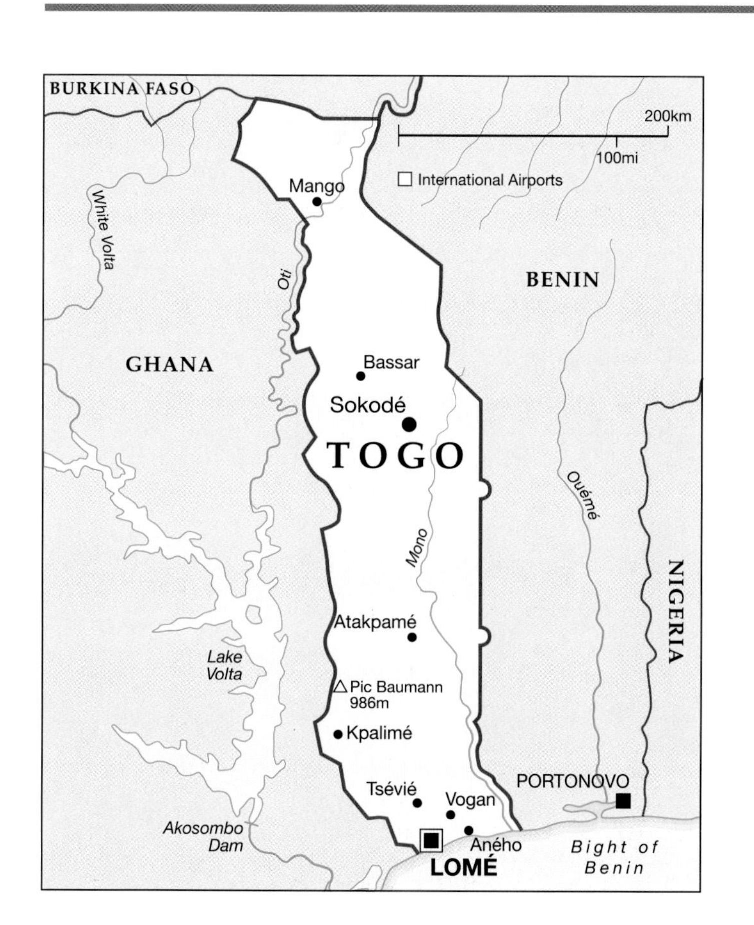

TOGO

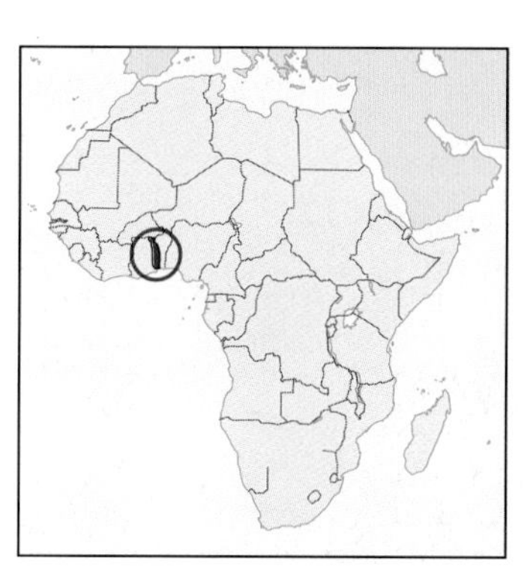

Facts and Figures

Capital: Lomé 500,000 (1987)
Area (sq km): 56,790
Population: 4,100,000 (1991)
Language: French, Ewe, Kabiye
Religion: Traditional 50%
Roman Catholic 26%
Muslim 15% Protestant 9%
Currency: CFA Franc = 100 centimes
Annual Income per person: $400
Annual Trade per person: $210
Adult Literacy: 43%
Life Expectancy (F): 57
Life Expectancy (M): 53

TUNISIA

Facts and Figures

Capital: Tunis 687,000 (1992)
Area (sq km): 164,150
Population: 8,370,000 (1992)
Language: Arabic, French
Religion: Muslim 97%
Currency: Tunisian Dinar = 1000 millimes
Annual Income per person: $1,510
Annual Trade per person: $1,065
Adult Literacy: 65%
Life Expectancy (F): 69
Life Expectancy (M): 67

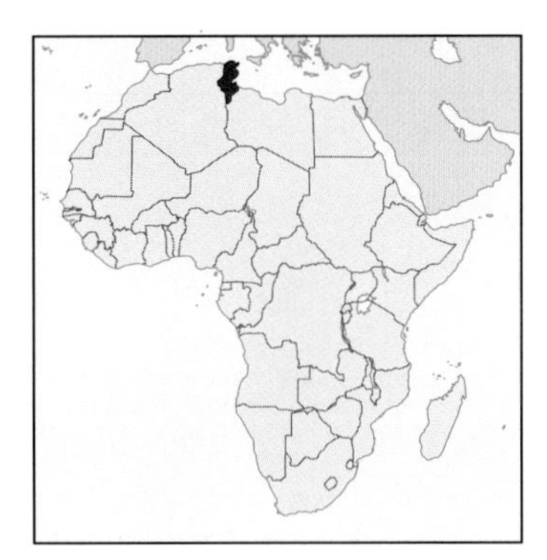

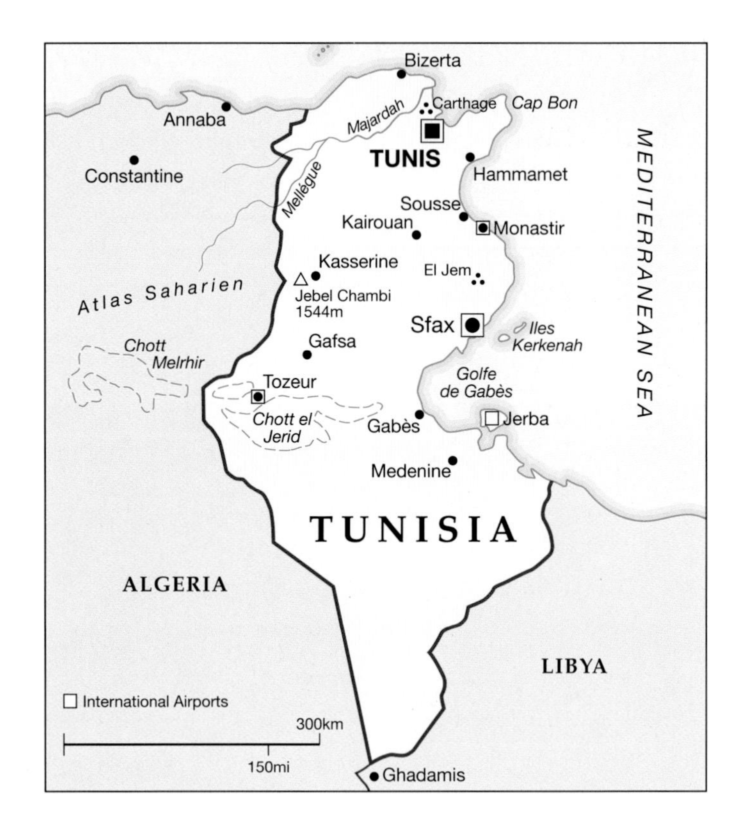

UGANDA

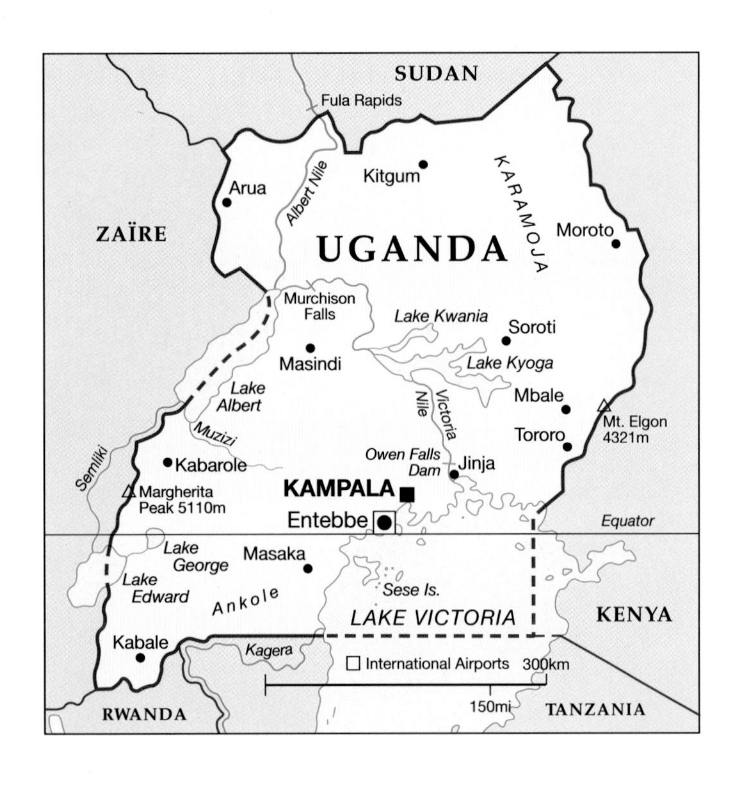

Facts and Figures

Capital: Kampala 773,500 (1991)
Area (sq km): 241,040
Population: 19,600,000 (1991)
Language: English, Swahili, Bantu languages
Religion: Roman Catholic 40%
Protestant 29%
Traditional 18% Muslim 7%
Currency: Uganda Shilling = 100 cents
Annual Income per person: $170
Annual Trade per person: $20
Adult Literacy: 48%
Life Expectancy (F): 55
Life Expectancy (M): 51

ZAIRE

Facts and Figures

Capital: Kinshasa 3,330,000 (1992)
Area (sq km): 2,344,890
Population: 40,256,000 (1991)
Language: French, Swahili, Tshiluba, Kikongo, Lingala
Religion: Roman Catholic 47%
Protestant 28%
Kimbanguiste 17%
Currency: Zaïre = 100 makuta
Annual Income per person: $220
Annual Trade per person: $42
Adult Literacy: 72%
Life Expectancy (F): 56
Life Expectancy (M): 52

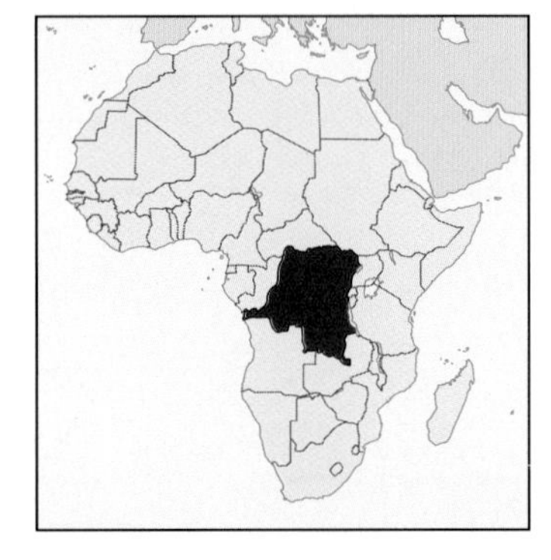

ZAMBIA

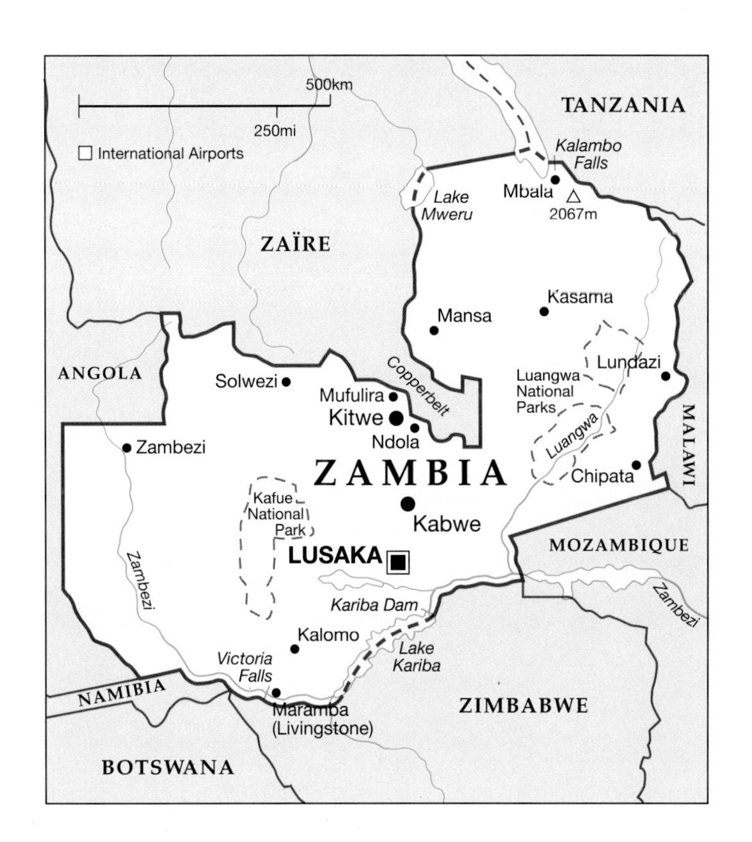

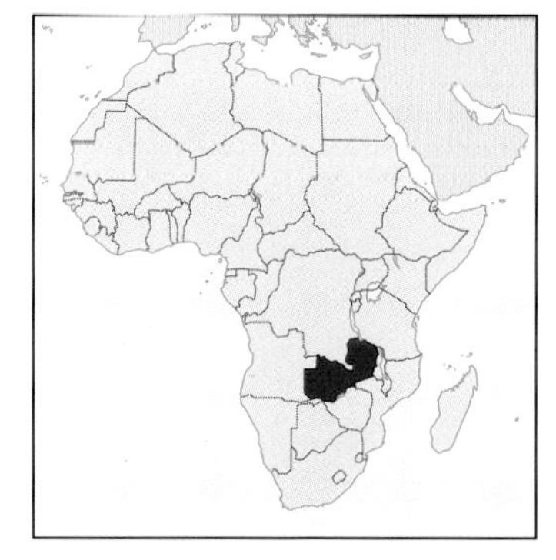

Facts and Figures

Capital: Lusaka 950,000 (1992)
Area (sq km): 752,610
Population: 8,780,000 (1991)
Language: English, Bantu languages
Religion: Christian 67% Traditional
Currency: Kwacha = 100 negwee
Annual Income per person: $420
Annual Trade per person: $265
Adult Literacy: 73%
Life Expectancy (F): 57
Life Expectancy (M): 54

ZIMBABWE

Facts and Figures

Capital: Harare 850,000 (1992)
Area (sq km): 390,000
Population: 10,700,000 (1992)
Language: English, Shona, Ndebele
Religion: Christian 53% Traditional
Currency: Zimbabwe Dollar = 100 cents
Annual Income per person: $620
Annual Trade per person: $381
Adult Literacy: 67%
Life Expectancy (F): 63
Life Expectancy (M): 59

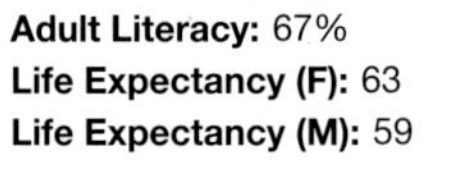

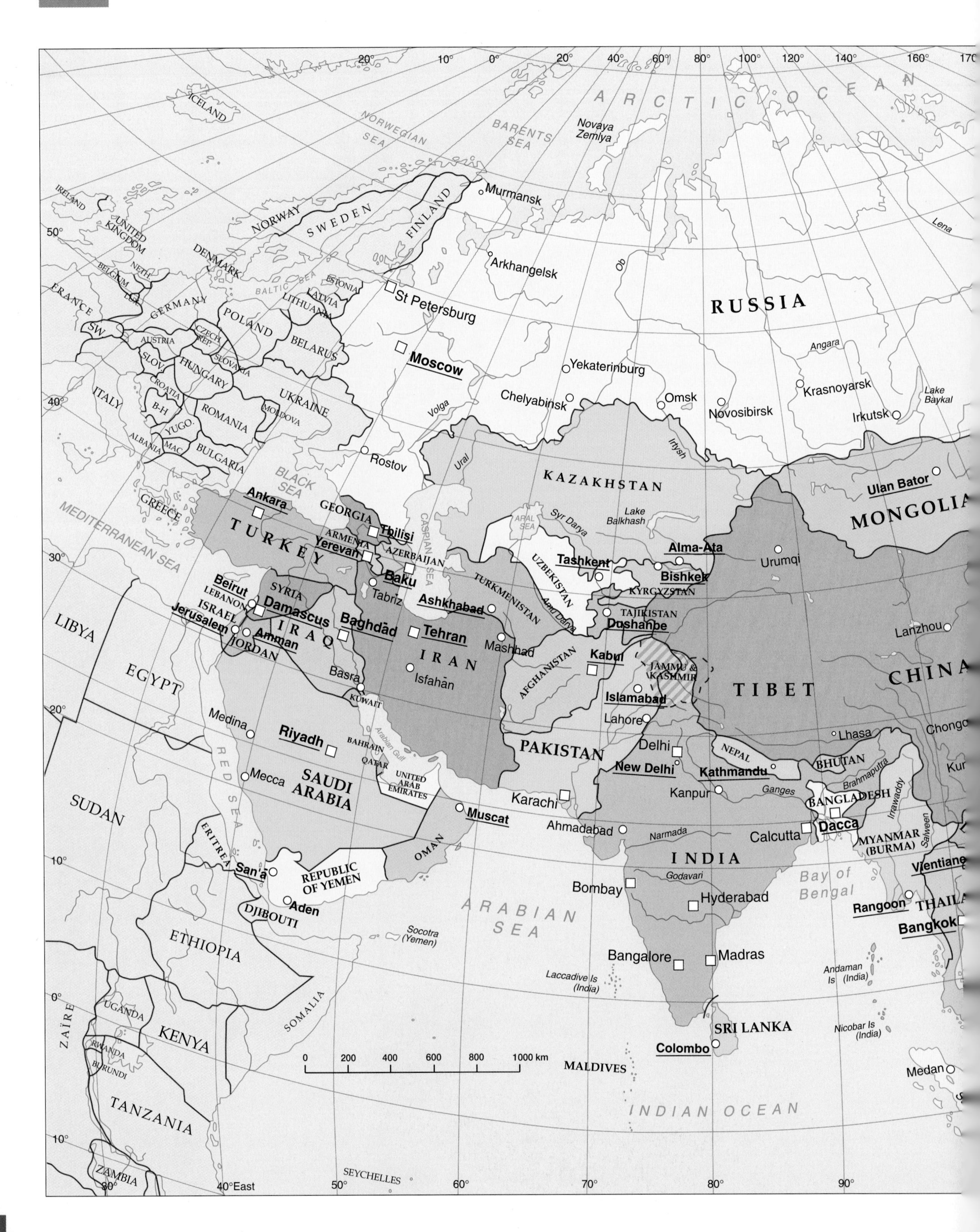
ARCTIC OCEAN
NORWEGIAN SEA
BARENTS SEA
Novaya Zemlya
ICELAND
IRELAND
UNITED KINGDOM
NORWAY
SWEDEN
FINLAND
Murmansk
DENMARK
NETH.
BELGIUM
LUX.
FRANCE
GERMANY
BALTIC SEA
ESTONIA
LATVIA
LITHUANIA
St Petersburg
Arkhangelsk
Ob
RUSSIA
Lena
SW
AUSTRIA
CZECH REP.
POLAND
BELARUS
SLOV.
SLOVAKIA
HUNGARY
Moscow
Yekaterinburg
Angara
ITALY
CROATIA
B-H
ROMANIA
MOLDOVA
UKRAINE
Volga
Chelyabinsk
Omsk
Novosibirsk
Krasnoyarsk
Irkutsk
Lake Baykal
YUGO.
ALBANIA
MAC
BULGARIA
Rostov
Ural
Irtysh
KAZAKHSTAN
Ulan Bator
MONGOLIA
GREECE
BLACK SEA
Ankara
GEORGIA
Tbilisi
TURKEY
ARMENIA
Yerevan
AZERBAIJAN
CASPIAN SEA
ARAL SEA
Syr Darya
Lake Balkhash
MEDITERRANEAN SEA
UZBEKISTAN
Tashkent
Alma-Ata
Bishkek
KYRGYZSTAN
Urumqi
Beirut
LEBANON
SYRIA
Baku
Tabriz
TURKMENISTAN
Ashkhabad
Amu Darya
TAJIKISTAN
Dushanbe
ISRAEL
Damascus
Jerusalem
Amman
JORDAN
IRAQ
Baghdād
Tehran
Mashhad
LIBYA
Lanzhou
EGYPT
IRAN
Isfahān
Basra
KUWAIT
AFGHANISTAN
Kabul
JAMMU & KASHMIR
Islamabad
TIBET
CHINA
Medina
Riyadh
BAHRAIN
Arabian Gulf
QATAR
Lahore
Lhasa
Chongq
PAKISTAN
Delhi
New Delhi
NEPAL
Kathmandu
BHUTAN
RED SEA
Mecca
SAUDI ARABIA
UNITED ARAB EMIRATES
Brahmaputra
Ganges
Kanpur
SUDAN
Karachi
Muscat
BANGLADESH
Irrawaddy
Salween
Ahmadabad
Narmada
Calcutta
Dacca
MYANMAR (BURMA)
ERITREA
OMAN
INDIA
San'a
REPUBLIC OF YEMEN
Godavari
Vientiane
Bay of Bengal
Bombay
Hyderabad
Rangoon
THAIL
Aden
DJIBOUTI
ARABIAN SEA
Socotra (Yemen)
Bangkok
ETHIOPIA
Bangalore
Madras
Andaman Is (India)
Laccadive Is (India)
UGANDA
SOMALIA
ZAÏRE
KENYA
SRI LANKA
Nicobar Is (India)
Colombo
RWANDA
BURUNDI
0 200 400 600 800 1000 km
MALDIVES
Medan
INDIAN OCEAN
TANZANIA
ZAMBIA
SEYCHELLES
30° 40°East 50° 60° 70° 80° 90°
20° 10° 0° 20° 40° 60° 80° 100° 120° 140° 160° 170
50° 40° 30° 20° 10° 0° 10°

COUNTRY	PAGE
Afghanistan	70
Armenia	70
Azerbaijan	70
Bahrain	71
Bangladesh	71
Bhutan	71
Brunei	72
Cambodia	72
China	73
Georgia	72
Hong Kong	73
India	74
Indonesia	74
Iran	74
Iraq	75
Israel	75
Japan	75
Jordan	76
North Korea	76
South Korea	76
Kazakhstan	77
Kuwait	77
Kyrgyzstan	77
Laos	78
Lebanon	78
Malaysia	78
Maldives	79
Mongolia	79
Myanmar (Burma)	79
Nepal	80
Oman	80
Pakistan	80
Philippines	81
Qatar	81
Russia	82
Saudi Arabia	82
Singapore	83
Sri Lanka	83
Syria	83
Taiwan	84
Tajikistan	84
Thailand	84
Turkmenistan	84
United Arab Emirates	85
Uzbekistan	85
Vietnam	85
Yemen	85

AFGHANISTAN

Facts and Figures

Capital: Kabul 913,200 (1979)
Area (sq km): 652,090
Population: 16,560,000 (1990)
Language: Pushtu, Dari
Religion: Sunni Muslim 70%
Shiite Muslim 25%
Currency: Afghani = 100 puls
Annual Income per person: $450
Annual Trade per person: $73
Adult Literacy: 30%
Life Expectancy (F): 44
Life Expectancy (M): 43

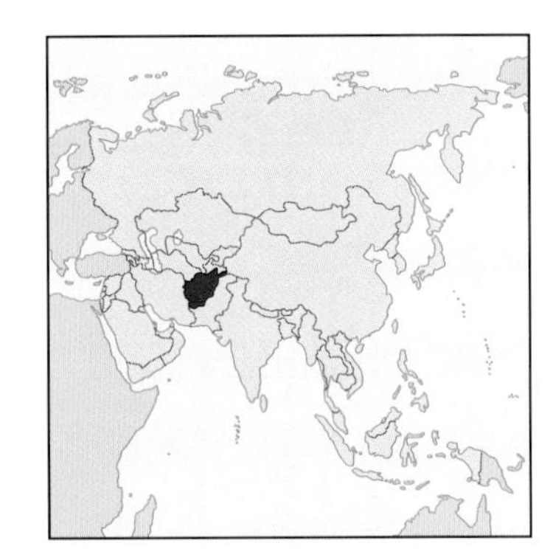

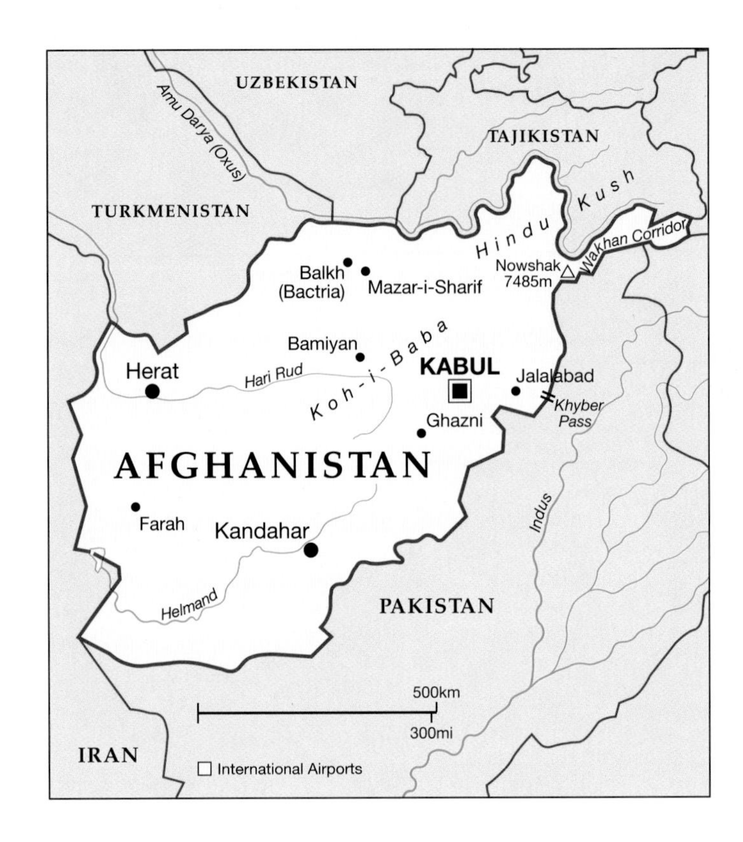

ARMENIA

Facts and Figures

Capital: Yerevan 1,185,000 (1990)
Area (sq km): 29,800
Population: 3,400,000 (1992)
Language: Armenian
Religion: Armenian Orthodox
Currency: Dram
Annual Income per person: $2,150
Annual Trade per person: $500
Adult Literacy: 93%
Life Expectancy (F): 73
Life Expectancy (M): 78

AZERBAIJAN

Facts and Figures

Capital: Baku 1,149,000 (1990)
Area (sq km): 86,600
Population: 7,200,000 (1992)
Language: Azerbaijani
Religion: Shiite Muslim 62%
Sunni Muslim 26%
Currency: Manat
Annual Income per person: $1,670
Annual Trade per person: $400
Adult Literacy: 93%
Life Expectancy (F): 75
Life Expectancy (M): 67

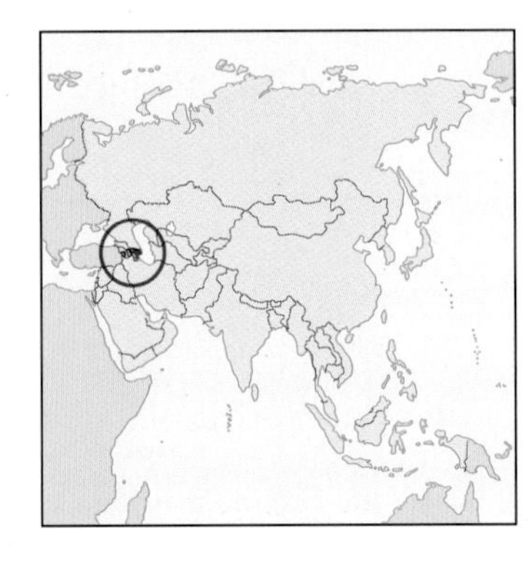

Dhahran
Al Khobar
SAUDI ARABIA
Causeway
Umm Nasan I.
Muharraq
Bahrain
MANAMA
Muharraq I.
Mina Sulman
Sitra I.
Awali
BAHRAIN
GULF OF BAHRAIN
Ra's al Bahr
Hawar Is. (in dispute)
☐ International Airports
20km
10mi
QATAR

BAHRAIN

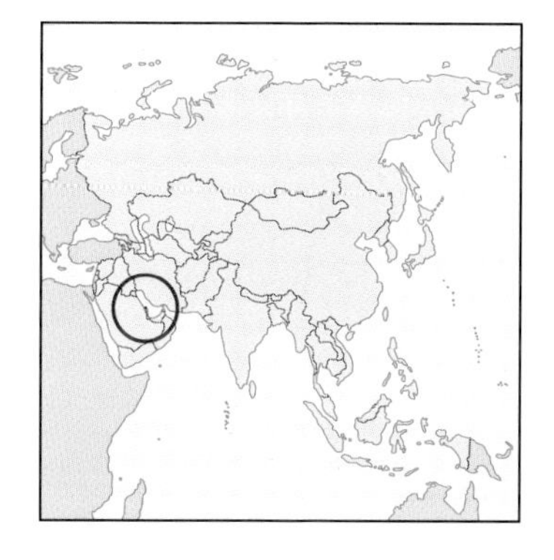

Facts and Figures

Capital: Manama 152,000 (1988)
Area (sq km): 688
Population: 538,000 (1993)
Language: Arabic
Religion: Shiite Muslim 60%
Sunni Muslim 25% Christian 7%
Currency: Bahrain Dinar = 1000 fils
Annual Income per person: $6,910
Annual Trade per person: $14,094
Adult Literacy: 77%
Life Expectancy (F): 74
Life Expectancy (M): 70

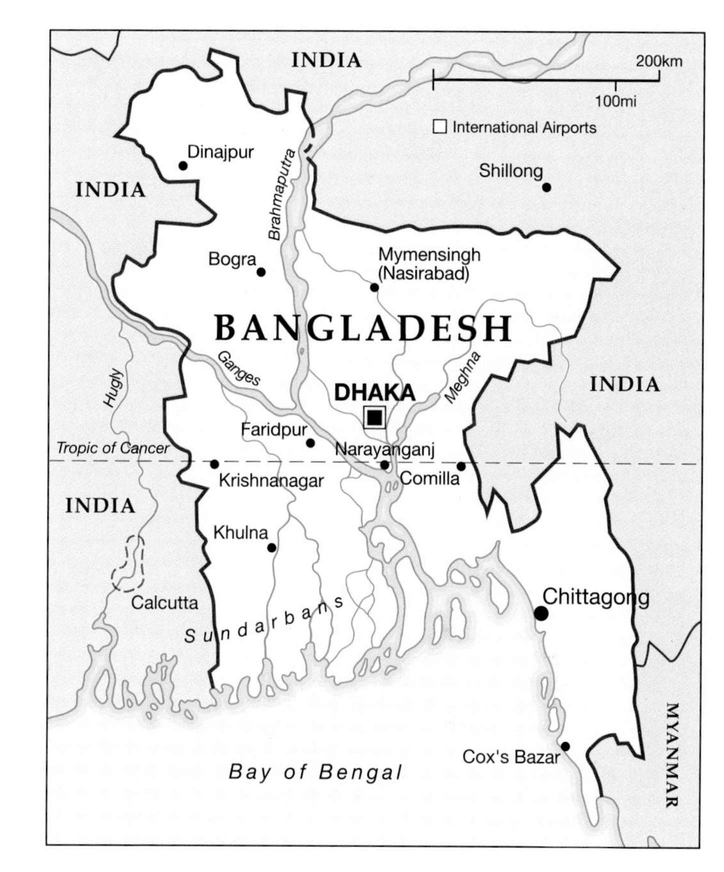

BANGLADESH

Facts and Figures

Capital: Dhaka 3,397,200 (1991)
Area (sq km): 148,400
Population: 118,700,000 (1993)
Language: Bengali, English
Religion: Sunni Muslim 85% Hindu 12%
Currency: Taka = 100 paisa
Annual Income per person: $220
Annual Trade per person: $43
Adult Literacy: 35%
Life Expectancy (F): 53
Life Expectancy (M): 53

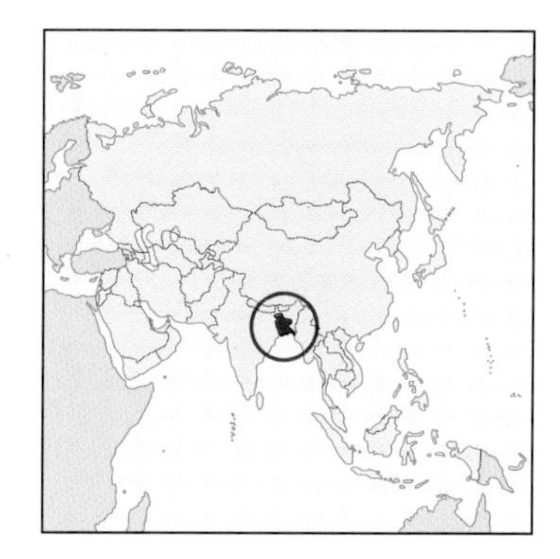

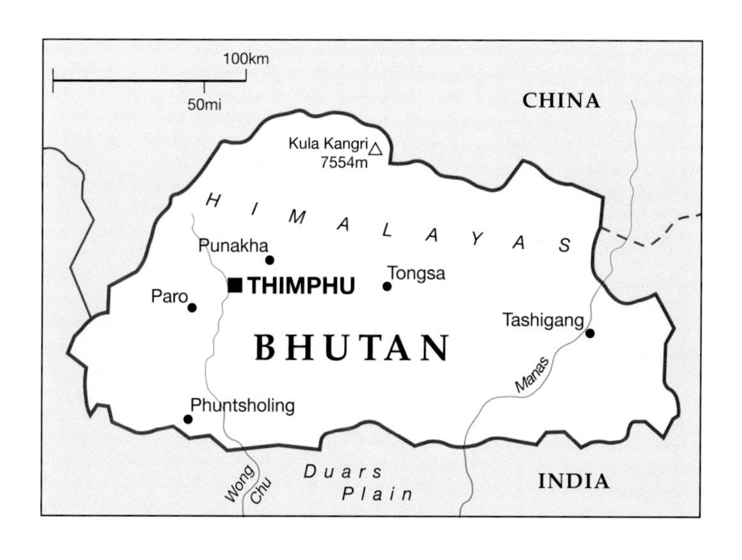

BHUTAN

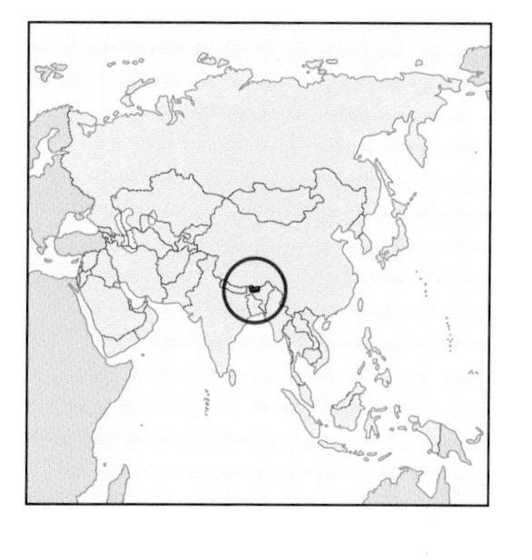

Facts and Figures

Capital: Thimphu 30,340 (1993)
Area (sq km): 46,500
Population: 600,000 (1990)
Language: Tibetan, Nepalese
Religion: Buddhist 65% Hindu 33%
Currency: Ngultrum = 100 chetrum
Annual Income per person: $468
Annual Trade per person: N/A
Adult Literacy: 35%
Life Expectancy (F): 49
Life Expectancy (M): 51

BRUNEI

Facts and Figures

Capital: Bandar Seri Begawan 45,870 (1991)
Area (sq km): 5,770
Population: 267,800 (1992)
Language: Malay, English, Chinese
Religion: Muslim 67% Buddhist 13%
Christian10%
Currency: Brunei Dollar
Annual Income per person: $15,390
Annual Trade per person: $11,108
Adult Literacy: 86%
Life Expectancy (F): 77
Life Expectancy (M): 74

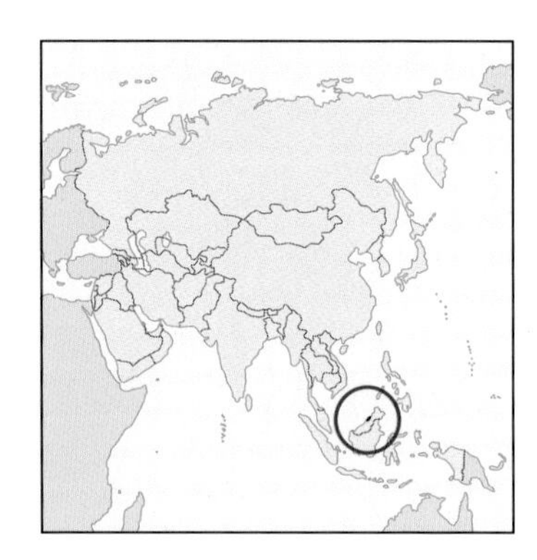

CAMBODIA

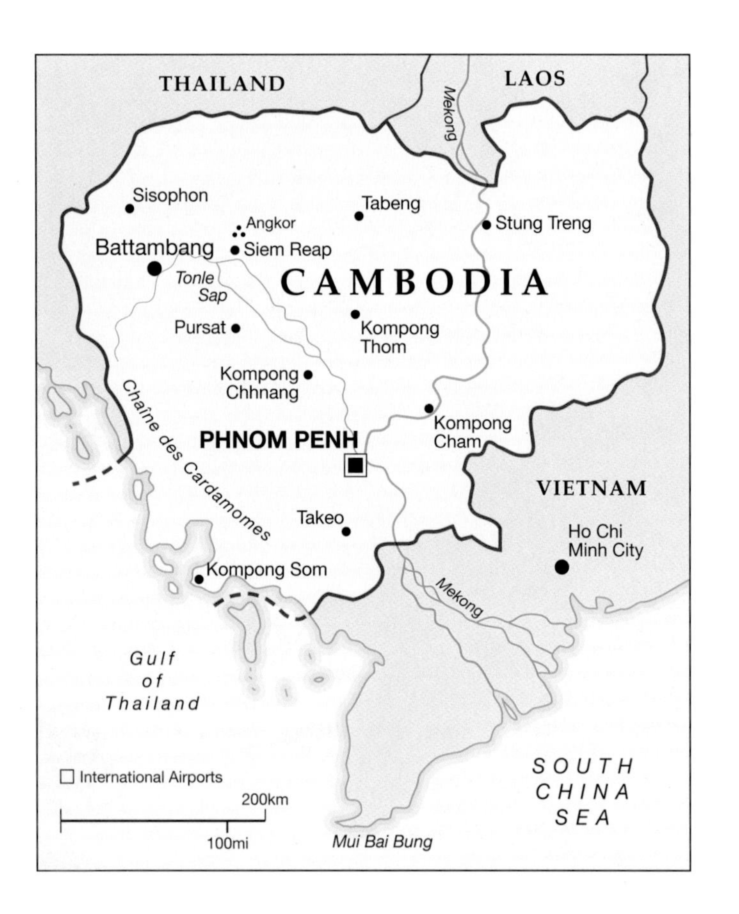

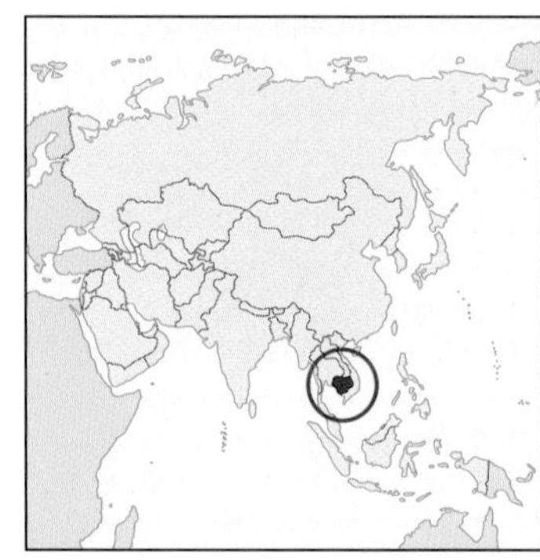

Facts and Figures

Capital: Phnom Penh 800,000 (1989)
Area (sq km): 181,040
Population: 12,000,000 (1993)
Language: Khmer, French
Religion: Buddhist 88%
Currency: Riel = 100 sen
Annual Income per person: $200
Annual Trade per person: $25
Adult Literacy: 35%
Life Expectancy (F): 52
Life Expectancy (M): 50

GEORGIA

Facts and Figures

Capital: Tbilisi 1,283,000(1991)
Area (sq km): 69,700
Population: 5,460,000 (1990)
Language: Georgian, Armenian, Russian
Religion: Orthodox 83% Muslim 11%
Currency: Rouble = 100 kopeks
Annual Income per person: $1,640
Annual Trade per person: $300
Adult Literacy: Data not available
Life Expectancy (F): 76
Life Expectancy (M): 69

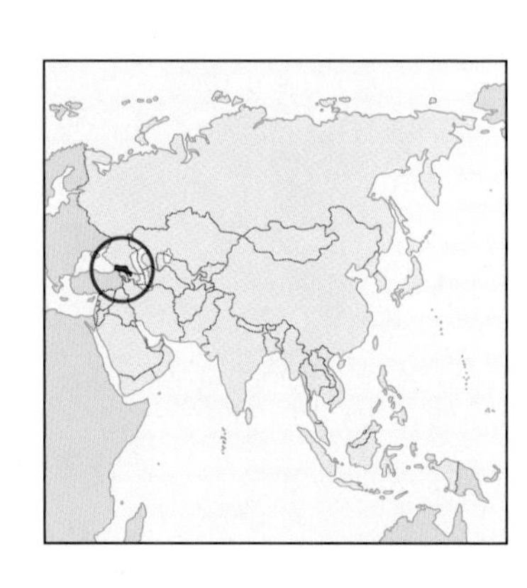

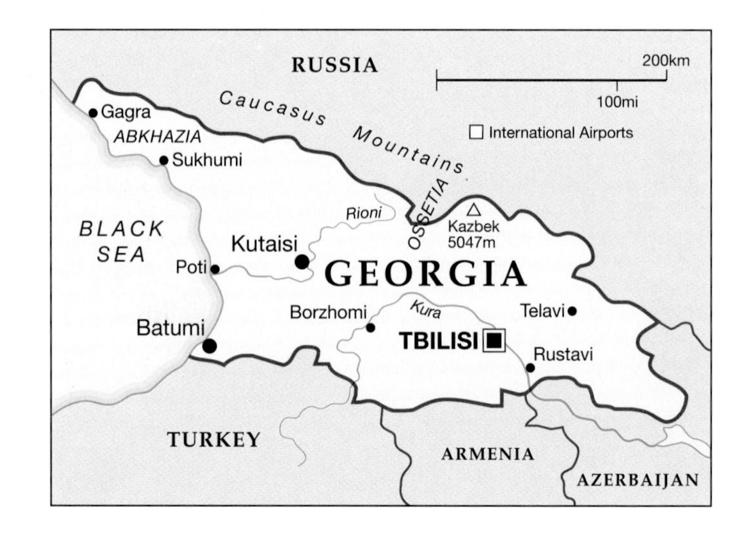

CHINA

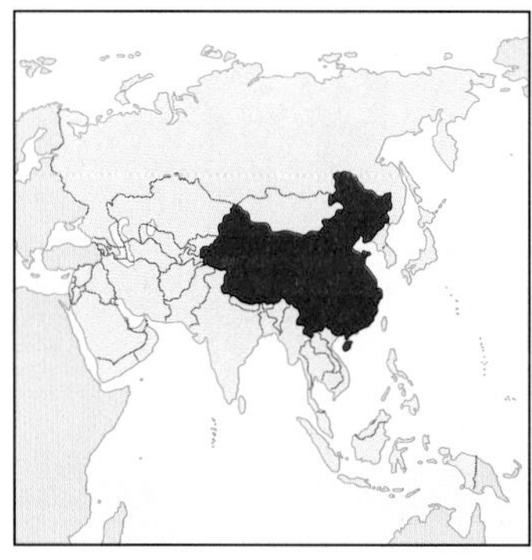

Facts and Figures

Capital: Beijing 7,050,000 (1992)
Area (sq km): 9,560,000
Population: 1,158,000,000 (1992)
Language: Chinese
Religion: Confucian 20% Buddhist 6% Taoist 2% Muslim 2%
Currency: Renminbi yuan = 10 jiao = 100 fen
Annual Income per person: $370
Annual Trade per person: $137
Adult Literacy: 70%
Life Expectancy (F): 73
Life Expectancy (M): 69

HONG KONG

Facts and Figures

British Colony until 1997
Area (sq km): 1,040
Population: 5,900,000
Language: English, Chinese
Religion: Buddist, Confucian, Taoist, Christian
Currency: Hong Kong Dollar
Annual Income per person: $13,200
Annual Trade per person: $41,886
Adult Literacy: 90%
Life Expectancy (F): 80
Life Expectancy (M): 75

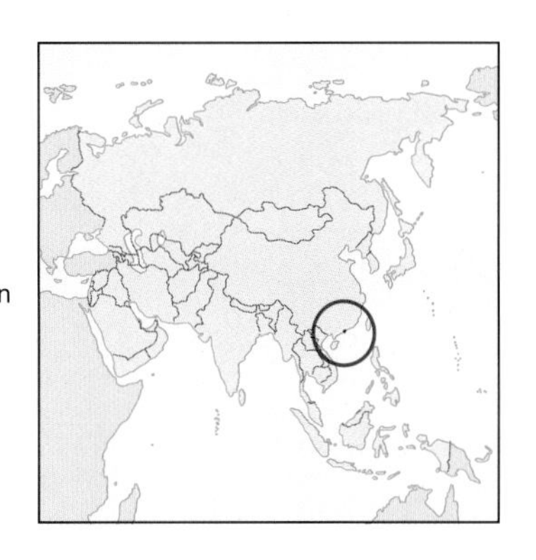

HONG KONG

INDIA

Facts and Figures

Capital: New Delhi 7,508,000 (1991)
Area (sq km): 3,165,600
Population: 903,000,000 (1991)
Language: Hindi, English, Teluga, Bengali, Marati, Urdu
Religion: Hindu 83% Muslim 11% Christian 2% Sikh 2% Buddhist 1%
Currency: Rupee = 100 paisa
Annual Income per person: $330
Annual Trade per person: $44
Adult Literacy: 48%
Life Expectancy (F): 61
Life Expectancy (M): 60

INDONESIA

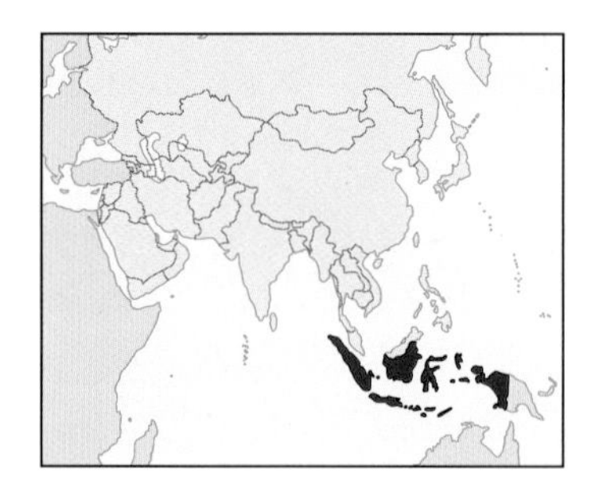

Facts and Figures

Capital: Jakarta 9,000,000 (1993)
Area (sq km): 1,919,440
Population: 187,800,000 (1993)
Language: Bahasa Indonesian, Dutch
Religion: Sunni Muslim 87% Christian 9%
Currency: Rupiah = 100 sen
Annual Income per person: $620
Annual Trade per person: $293
Adult Literacy: 76%
Life Expectancy (F): 65
Life Expectancy (M): 61

IRAN

Facts and Figures

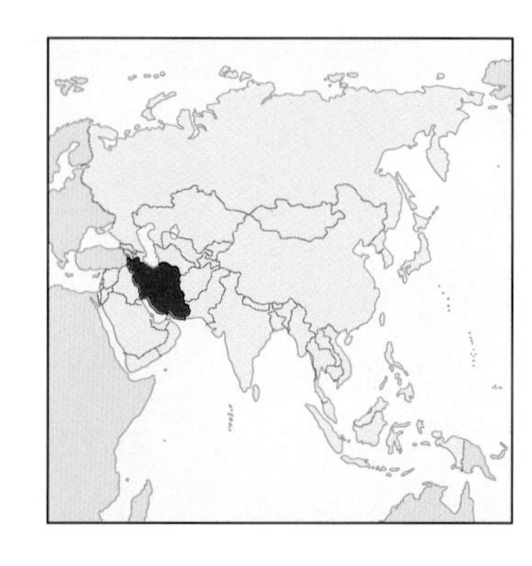

Capital: Tehran 7,214,000 (1992)
Area (sq km): 1,648,000
Population: 61,600,000 (1992)
Language: Farsi, Kurdish, Baluchi
Religion: Shi'ite Muslim 94%
Currency: Rial = 100 dinars
Annual Income per person: $2,320
Annual Trade per person: $598
Adult Literacy: 54%
Life Expectancy (F): 68
Life Expectancy (M): 67

TURKEY
Lake Urmia
Kurdistan
Nineveh
Mosul
Arbil
Firat
SYRIA
Al Jazirah
Hatra
Tharthar
Tigris
Kirkuk
IRAN
Samarra
Euphrates
BAGHDAD
JORDAN
IRAQ
Ctesiphon
Babylon
Al Kut
Karun
An Najaf
SYRIAN DESERT
Ur
SAUDI ARABIA
Al Basrah
International Airports
400km
200mi
KUWAIT
The Gulf

IRAQ

Facts and Figures

Capital: Baghdad 4,914,000 (1992)
Area (sq km): 438,320
Population: 19,410,000 (1993)
Language: Arabic, Kurdish, Turkish
Religion: Shi'ite Muslim 61%
Sunni Muslim 34%
Currency: Dinar = 20 dirhams
Annual Income per person: $2,100
Annual Trade per person: $276
Adult Literacy: 60%
Life Expectancy (F): 67
Life Expectancy (M): 65

ISRAEL

Facts and Figures

Capital: Jerusalem 556,500 (1991)
Area (sq km): 27,010
Population: 7,200,000
Language: Hebrew, Arabic, English
Religion: Jewish 78% Muslim 13%
Christian 2%
Currency: Shekel = 100 agorot
Annual Income per person: $11,330
Annual Trade per person: $5,794
Adult Literacy: 96%
Life Expectancy (F): 78
Life Expectancy (M): 74

800km
400mi
RUSSIA
International Airports
Hokkaido
Sapporo
CHINA
Hakodate
Islands occupied by Russia
Aomori
NORTH KOREA
SEA OF JAPAN
Honshu
Sendai
Sado I.
Noto Peninsula
JAPAN
SOUTH KOREA
Oki Is.
Mt. Fuji 3776m
TOKYO
Boso Peninsula
Kyoto
Hiroshima
Kobe
Yokohama & Kawasaki
Nagoya
Osaka
Kitakyushu
Fukuoka
Takamatsu
Kii Peninsula
Nagasaki
Shikoku
Kyushu
EAST CHINA SEA
Osumi Is.
Tokara Is.
Ryukyu Islands
PACIFIC OCEAN
Amami Is.
Bonin Is.
Volcano Is.
Okinawa
Iwo Jima
Sakashima Is.

JAPAN

Facts and Figures

Capital: Tokyo 7,976,000 (1992)
Area (sq km): 377,750
Population: 124,900,000 (1992)
Language: Japanese, Korean, Chinese
Religion: Shintoism, Buddhism
Currency: Yen = 100 sen
Annual Income per person: $29,794
Annual Trade per person: $4,617
Adult Literacy: 99%
Life Expectancy (F): 82
Life Expectancy (M): 76

JORDAN

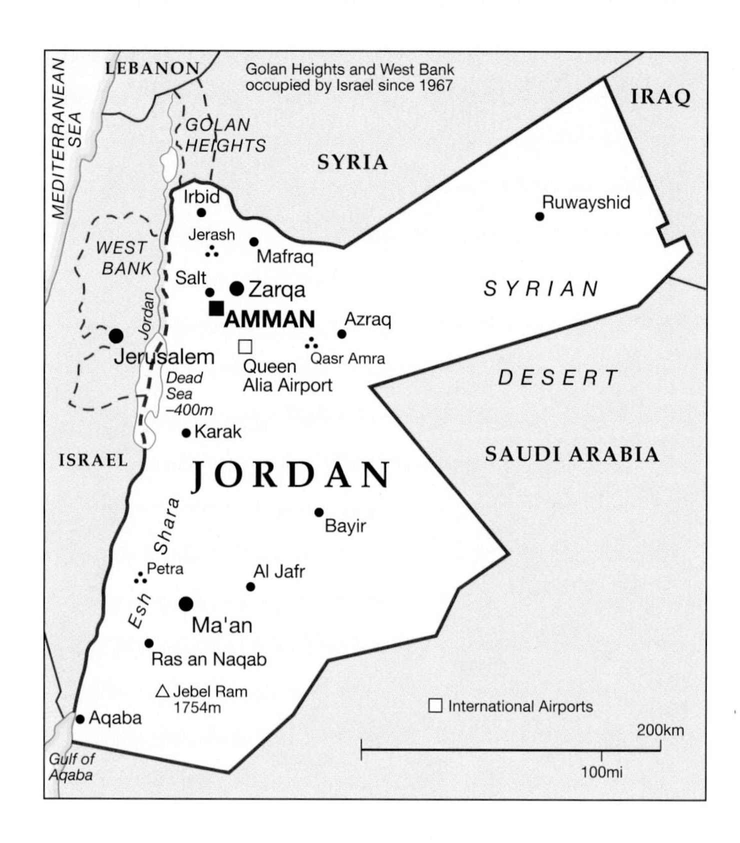

Facts and Figures

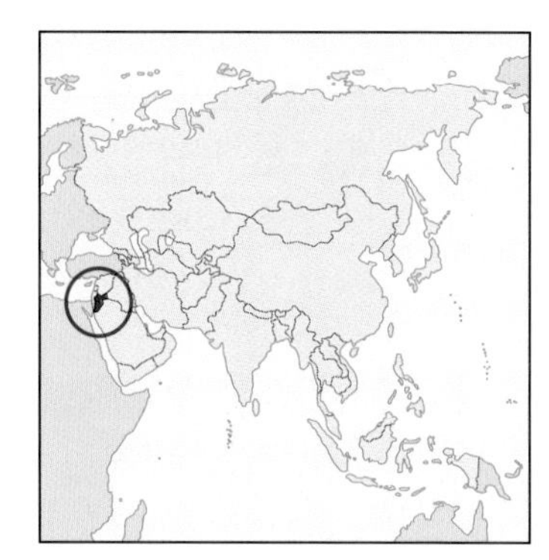

Capital: Amman 1,272,000 (1992)
Area (sq km): 97,740
Population: 4,010,000 (1992)
Language: Arabic
Religion: Sunni Muslim 95% Christian 5%
Currency: Jordanian Dinar = 1000 fils
Annual Income per person: $1,120
Annual Trade per person: $825
Adult Literacy: 80%
Life Expectancy (F): 70
Life Expectancy (M): 66

NORTH KOREA

Facts and Figures

Capital: Pyongyang 2,567,000 (1992)
Area (sq km): 122,762
Population: 22,030,000 (1991)
Language: Korean, Chinese
Religion: Chondogyo 14% Traditional 14% Buddhist 2% Christian 1%
Currency: North Korean Won = 100 chon
Annual Income per person: $1,100
Annual Trade per person: $200
Adult Literacy: 99%
Life Expectancy (F): 72
Life Expectancy (M): 66

SOUTH KOREA

Facts and Figures

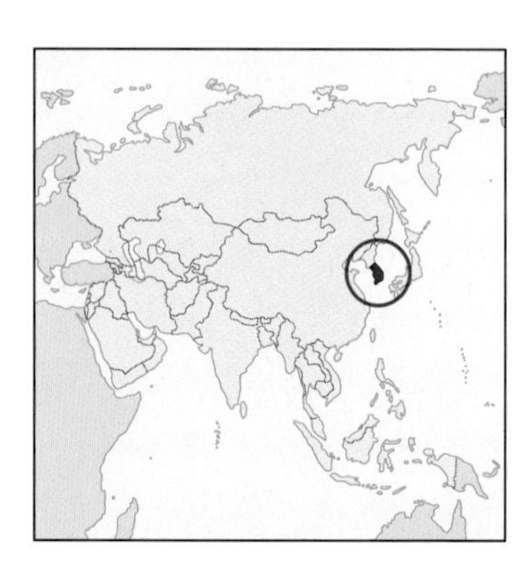

Capital: Seoul 10,268,000 (1990)
Area (sq km): 99,270
Population: 43,663,000 (1990)
Language: Korean
Religion: Buddhist 24% Protestant 16% RC 5% Confucian 2%
Currency: South Korean Won = 100 chon
Annual Income per person: $6,743
Annual Trade per person: $3,546
Adult Literacy: 96%
Life Expectancy (F): 73
Life Expectancy (M): 67

KAZAKHSTAN

Facts and Figures

Capital: Alma-Ata 1,166,000 (1992)
Area (sq km): 2,717,300
Population: 17,200,000 (1992)
Language: Kazakh, Russian
Religion: Sunni Muslim, Christian
Currency: Tenge = 500 rubles
Annual Income per person: $2,470
Annual Trade per person: $500
Adult Literacy: Data not available
Life Expectancy (F): 73
Life Expectancy (M): 64

KUWAIT

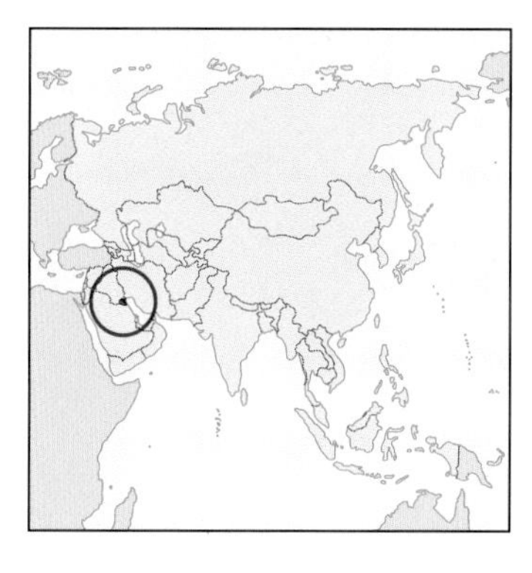

Facts and Figures

Capital: Kuwait City 31,300 (1993)
Area (sq km): 17,820
Population: 2,100,000 (1991)
Language: Arabic, English
Religion: Sunni Muslim 78%
Shiite Muslim 14% Christian 6%
Currency: Kuwaiti Dinar = 1000 fils
Annual Income per person: $16,380
Annual Trade per person: $8,682
Adult Literacy: 74%
Life Expectancy (F): 77
Life Expectancy (M): 72

KYRGYZSTAN

Facts and Figures

Capital: Bishkek 641,400 (1991)
Area (sq km): 198,500
Population: 4,500,000 (1992)
Language: Kirghiz
Religion: Sunni Muslim, Christian
Currency: Som = 100 tyiyn
Annual Income per person: $1,550
Annual Trade per person: $30
Adult Literacy: Data not available
Life Expectancy (F): 73
Life Expectancy (M): 65

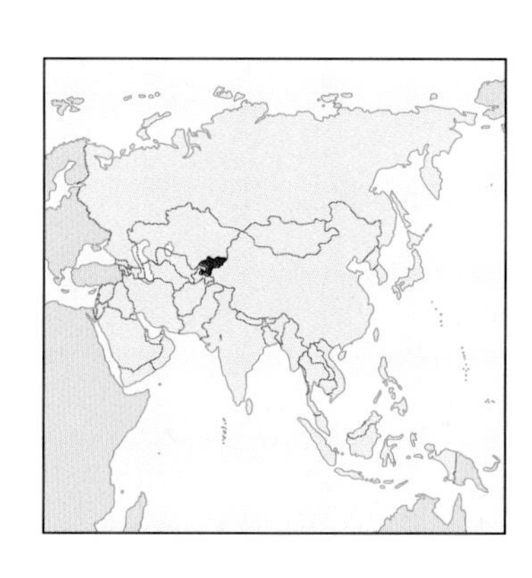

LAOS

Facts and Figures

Capital: Vientiane 454,000 (1992)
Area (sq km): 236,800
Population: 4,400,000 (1993)
Language: Lao, French, English
Religion: Buddhist 52% Traditional 33%
Currency: Kip = 100 cents
Annual Income per person: $230
Annual Trade per person: $62
Adult Literacy: 84%
Life Expectancy (F): 53
Life Expectancy (M): 50

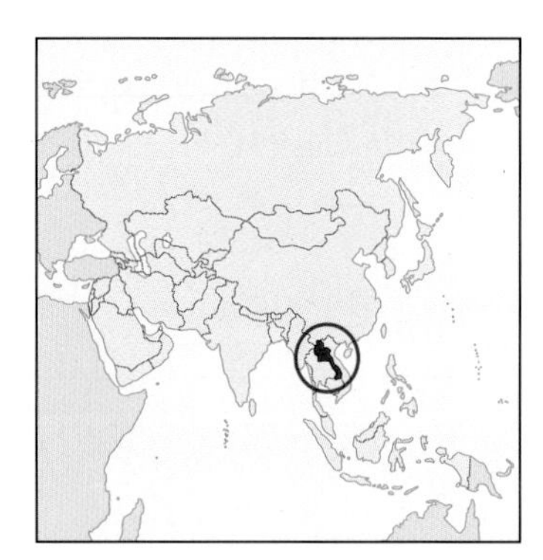

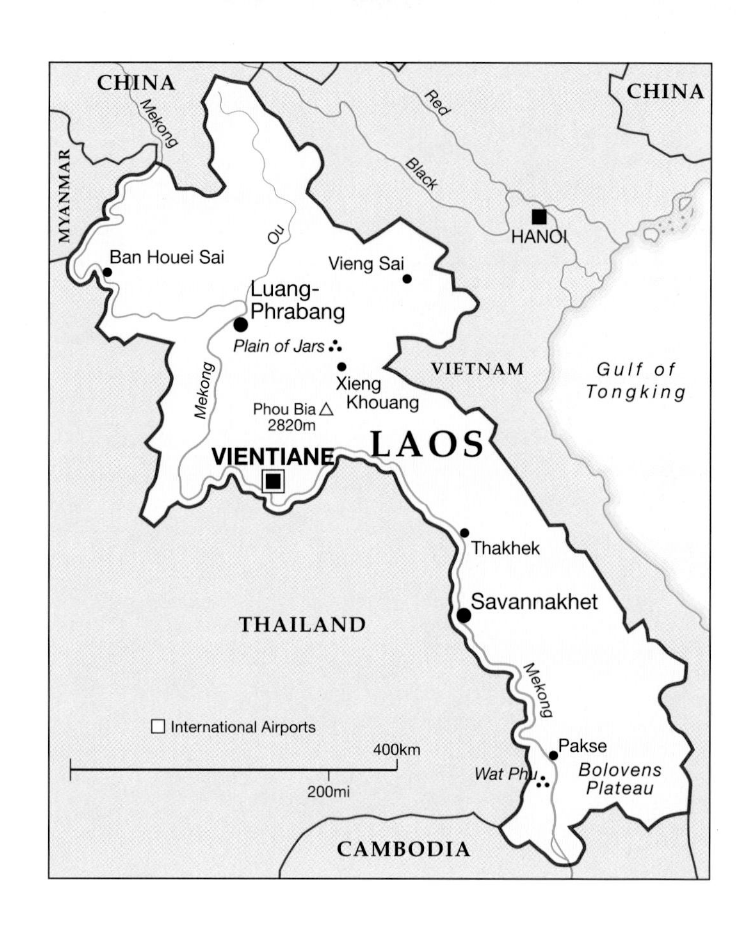

LEBANON

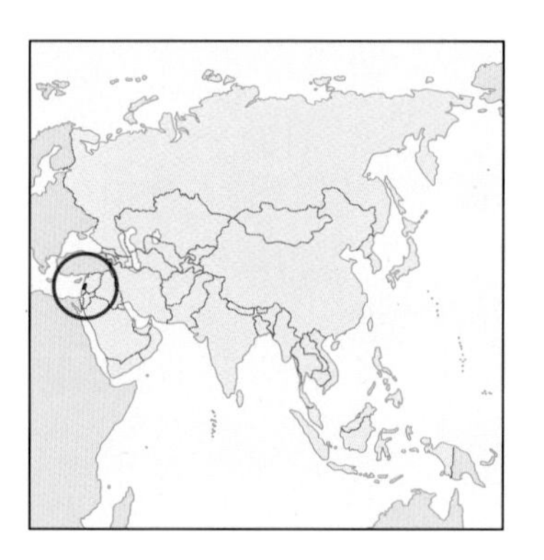

Facts and Figures

Capital: Beirut 1,500,000 (1992)
Area (sq km): 10,400
Population: 2,760,000 (1991)
Language: Arabic, French, English, Armenian
Religion: Shi'ite Muslim 35% Maronite Christian 27% Sunni Muslim 23%
Currency: Lebanese Pound = 100 piastres
Annual Income per person: $2,000
Annual Trade per person: $750
Adult Literacy: 80%
Life Expectancy (F): 69
Life Expectancy (M): 65

MALAYSIA

Facts and Figures

Capital: Kuala Lumpur 1,233,000 (1990)
Area (sq km): 329,760
Population: 19,030,000 (1993)
Language: Bahasa Malaysian, Chinese, Tamil, English
Religion: Sunni Muslim 53% Buddhist 17% Taoist, Hindu, Christian
Currency: Ringgit = 100 sen
Annual Income per person: $3,265
Annual Trade per person: $3,877
Adult Literacy: 79%
Life Expectancy (F): 73
Life Expectancy (M): 69

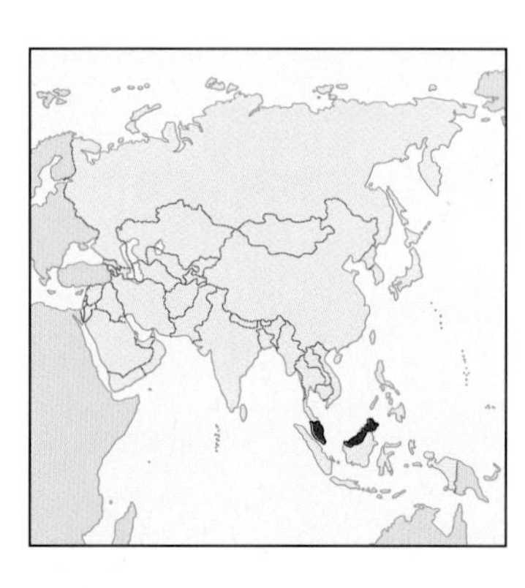

MALDIVES

INDIAN OCEAN
Haa Dhaal Atoll
Raa Atoll
Kuredhdhu
Lhaviyani Atoll
Kunfunadhoo
Kaafu Atoll
MALÉ
Alif Atoll
Dhiggiri
Athimatha
Thaa Atoll
MALDIVES
One and Half Degree Channel
Gaaf Alif Atoll
Equator
International Airports
Seena Atoll
Gan
200km
100mi

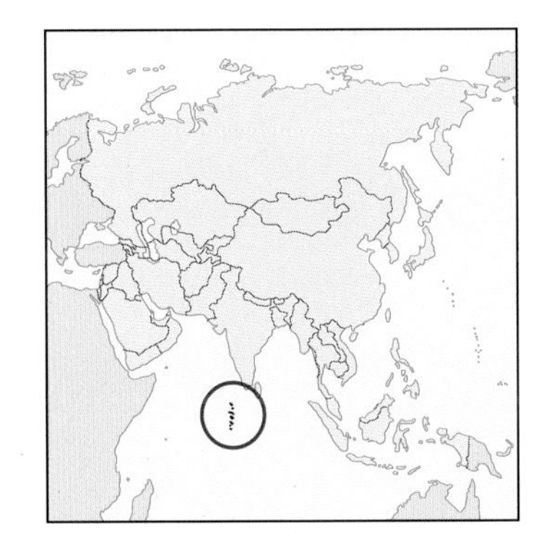

Facts and Figures

Capital: Malé 55,130 (1990)
Area (sq km): 298
Population: 238,400 (1993)
Language: Divehi, English, Arabic
Religion: Sunni Muslim
Currency: Rufiyaa = 100 laari
Annual Income per person: $460
Annual Trade per person: $750
Adult Literacy: 92%
Life Expectancy (F): 63
Life Expectancy (M): 61

MYANMAR (BURMA)

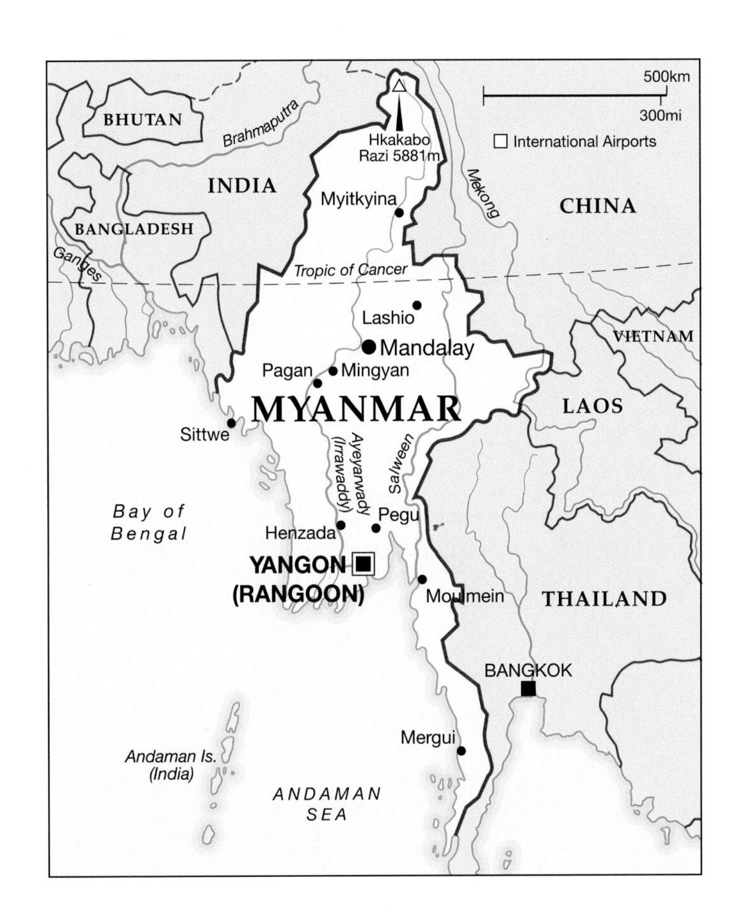

Facts and Figures

Capital: Rangoon 2,458,710 (1983)
Area (sq km): 676,577
Population: 42,330,000 (1993)
Language: Burmese, English
Religion: Buddhist
Currency: Kyat = 100 pyas
Annual Income per person: $500
Annual Trade per person: $24
Adult Literacy: 81%
Life Expectancy (F): 64
Life Expectancy (M): 61

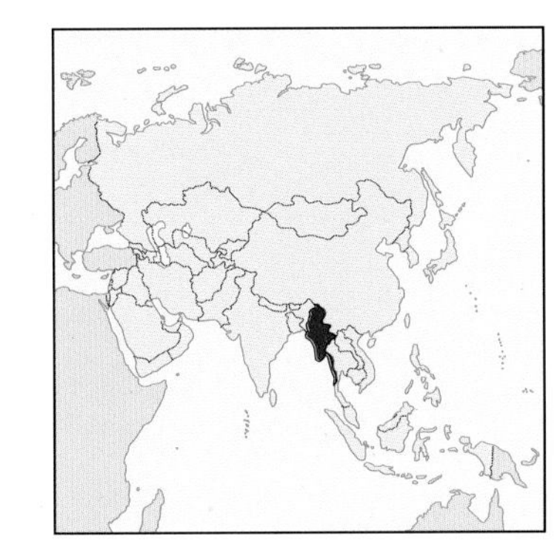

MONGOLIA

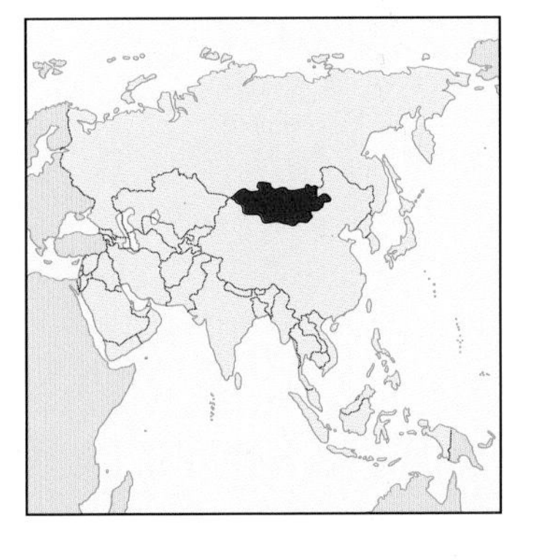

Facts and Figures

Capital: Ulan Bator 55,000 (1991)
Area (sq km): 1,566,500
Population: 2,260,000 (1992)
Language: Halh Mongol, Chinese, Russian
Religion: Shamanist 31% Muslim 4% Buddhist
Currency: Tugrik = 100 mongo
Annual Income per person: $600
Annual Trade per person: $400
Adult Literacy: 91%
Life Expectancy (F): 65
Life Expectancy (M): 62

NEPAL

Facts and Figures

Capital: Kathmandu 419,000 (1991)
Area (sq km): 147,200
Population: 19,360,000 (1991)
Language: Nepalese, Tibetan
Religion: Hindu 89% Buddhist 5%
Currency: Nepalese Rupee = 100 paisa
Annual Income per person: $180
Annual Trade per person: $54
Adult Literacy: 26%
Life Expectancy (F): 53
Life Expectancy (M): 54

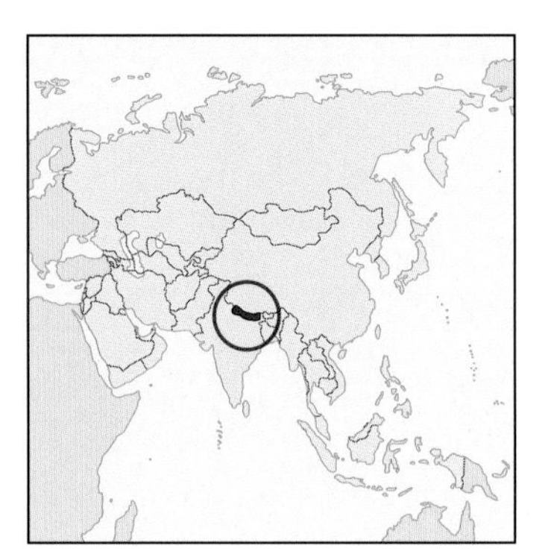

OMAN

Facts and Figures

Capital: Muscat 320,000 (1992)
Area (sq km): 309,500
Population: 2,070,000 (1991)
Language: Arabic, English, Baluchi
Religion: Muslim 86%
Currency: Omani Rial = 1000 baiza
Annual Income per person: $5,500
Annual Trade per person: $5,172
Adult Literacy: 35%
Life Expectancy (F): 70
Life Expectancy (M): 66

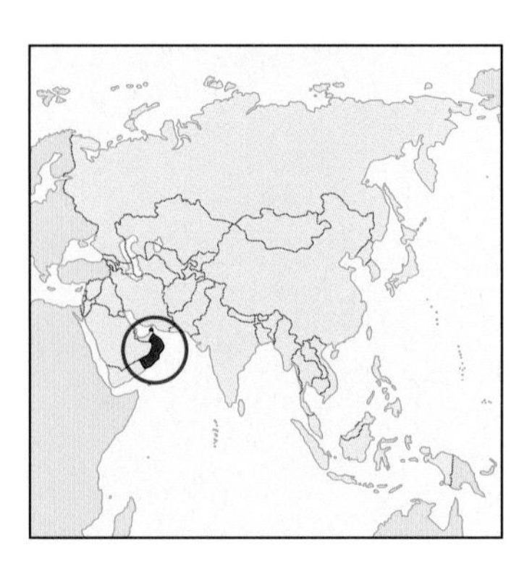

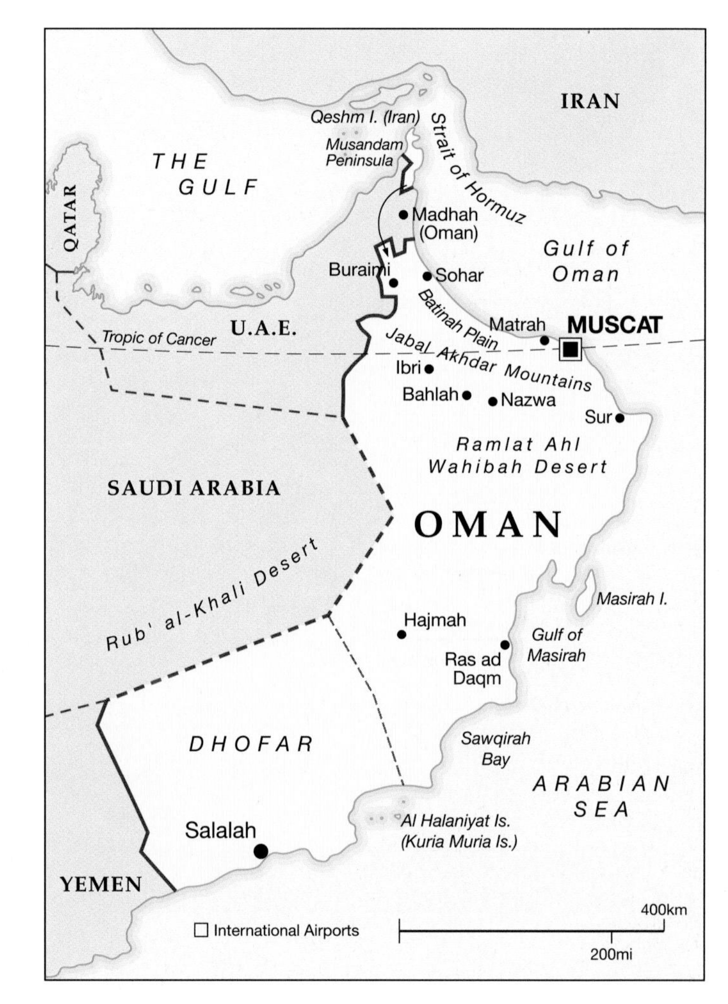

PAKISTAN

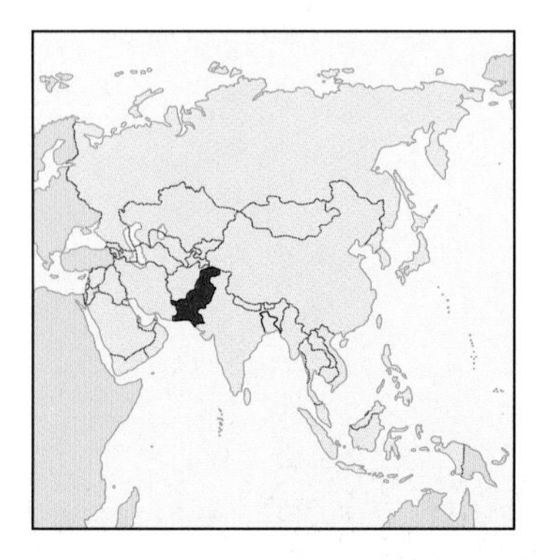

Facts and Figures

Capital: Islamabad 272,000 (1992)
Area (sq km): 796,100
Population: 122,400,000 (1992)
Language: Urdu, Punjabi, Sindhi, Baluchi, Pushtu, English
Religion: Muslim 96% Hindu, Christian
Currency: Pakistan Rupee = 100 paisa
Annual Income per person: $400
Annual Trade per person: $129
Adult Literacy: 35%
Life Expectancy (F): 59
Life Expectancy (M): 59

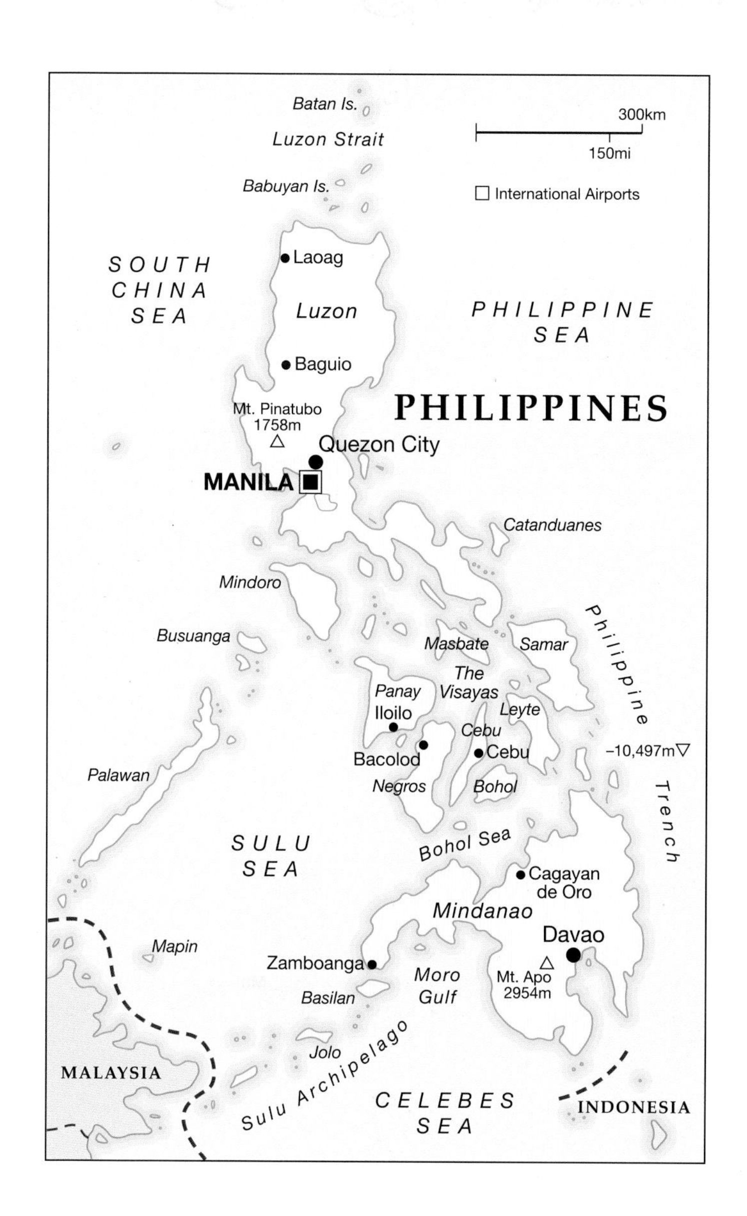

PHILIPPINES

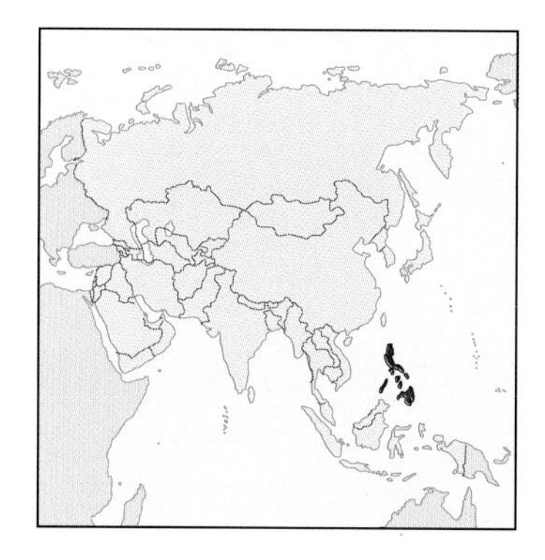

Facts and Figures

Capital: Manila 1,599,000 (1990)
Area (sq km): 300,000
Population: 65,650,000 (1993)
Language: Pilipino, English
Religion: RC 76% Protestant 5%
Sunni Muslim 4%
Currency: Piso = 100 sentimos
Annual Income per person: $740
Annual Trade per person: $332
Adult Literacy: 90%
Life Expectancy (F): 70
Life Expectancy (M): 63

QATAR

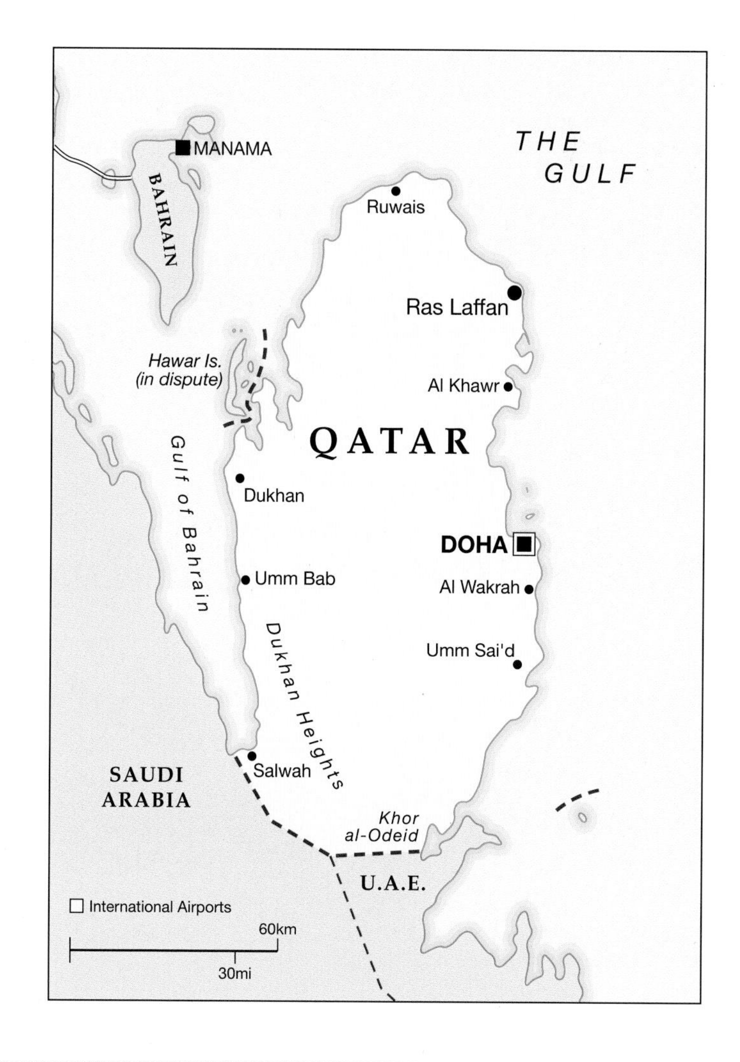

Facts and Figures

Capital: Doha 286,000 (1992)
Area (sq km): 11,500
Population: 510,000
Language: Arabic
Religion: Sunni Muslim 92%
Currency: Qatari Riyal = 100 dirhams
Annual Income per person: $15,870
Annual Trade per person: $28,296
Adult Literacy: 82%
Life Expectancy (F): 73
Life Expectancy (M): 68

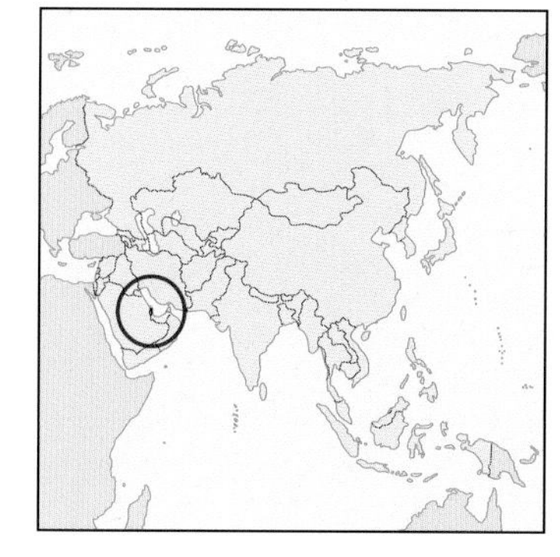

RUSSIA

Facts and Figures

Capital: Moscow 8,967,000 (1990)
Area (sq km): 17,075,000
Population: 148,700,000 (1992)
Language: Russian
Religion: Russian Orthodox 25%, Muslim, Buddhist
Currency: Rouble = 100 kopeks
Annual Income per person: $3,222
Annual Trade per person: $680
Adult Literacy: 94%
Life Expectancy (F): 74
Life Expectancy (M): 64

SAUDI ARABIA

Facts and Figures

Capital: Riyadh 2,320,000 (1992)
Area (sq km): 2,200,000
Population: 16,900,000 (1992)
Language: Arabic
Religion: Sunni Muslim 92%
Shi'ite Muslim 8%
Currency: Rial = 100 halalas
Annual Income per person: $7,070
Annual Trade per person: $4,606
Adult Literacy: 62%
Life Expectancy (F): 68
Life Expectancy (M): 64

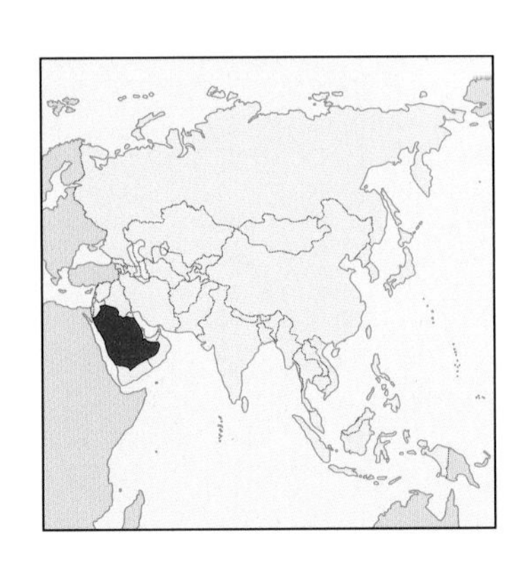

SINGAPORE

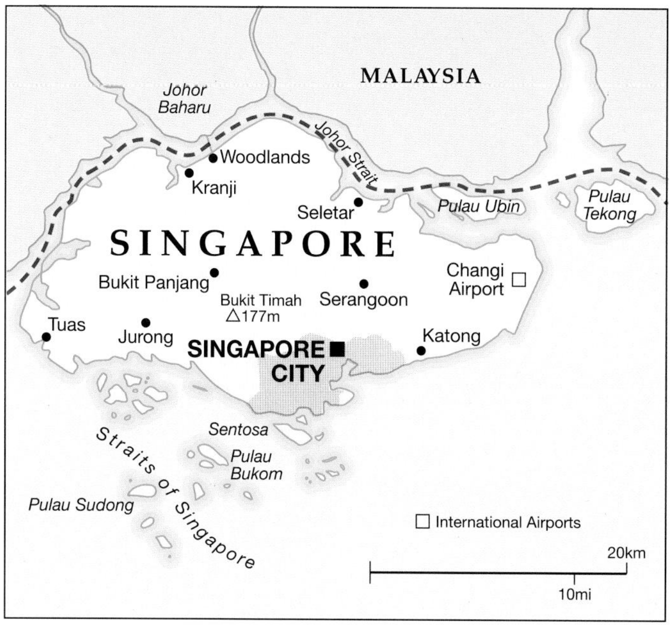

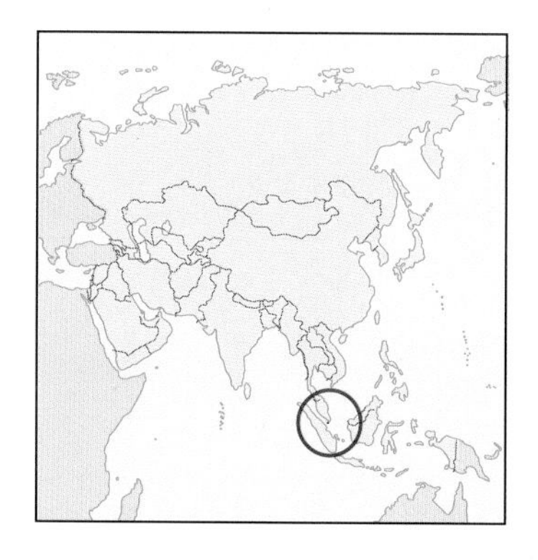

Facts and Figures

Capital: Singapore 2,820,000 (1992)
Area (sq km): 640
Population: 2,820,000 (1992)
Language: Chinese, Malay, Tamil, English
Religion: Buddhist 28% Taoist 13% Christian 19% Muslim 16% Hindu 5%
Currency: Singapore Dollar
Annual Income per person: $13,060
Annual Trade per person: $48,275
Adult Literacy: 88%
Life Expectancy (F): 77
Life Expectancy (M): 72

SRI LANKA

Facts and Figures

Capital: Colombo 615,000 (1990)
Area (sq km): 65,610
Population: 17,400,000 (1992)
Language: Sinhala, Tamil, English
Religion: Buddhist 64% Hindu 13% Muslim 8% Christian 7%
Currency: Sri Lankan Rupee = 100 cents
Annual Income per person: $539
Annual Trade per person: $293
Adult Literacy: 88%
Life Expectancy (F): 74
Life Expectancy (M): 70

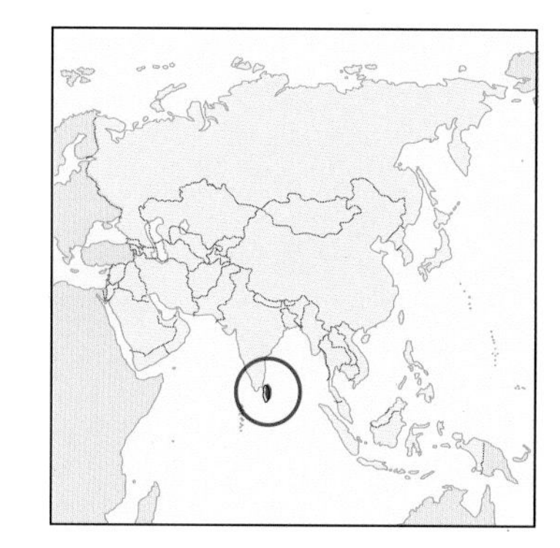

SYRIA

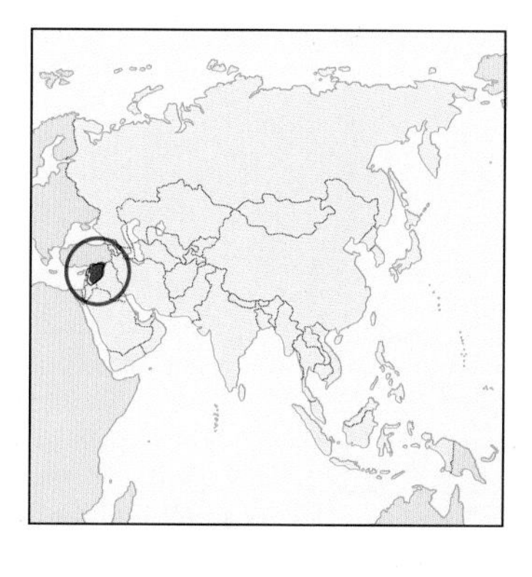

Facts and Figures

Capital: Damascus 1,450,000 (1993)
Area (sq km): 185,180
Population: 13,400,000 (1993)
Language: Arabic, Kurdish, Armenian
Religion: Sunni Muslim 90% Christian 9%
Currency: Syrian Pound = 100 piastres
Annual Income per person: $1,100
Annual Trade per person: $502
Adult Literacy: 64%
Life Expectancy (F): 69
Life Expectancy (M): 65

TAIWAN

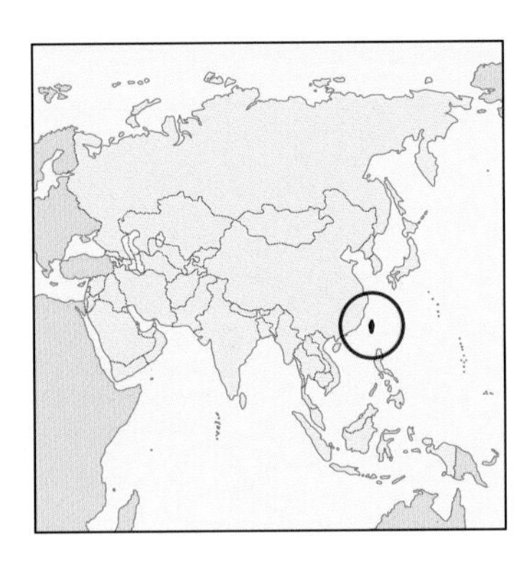

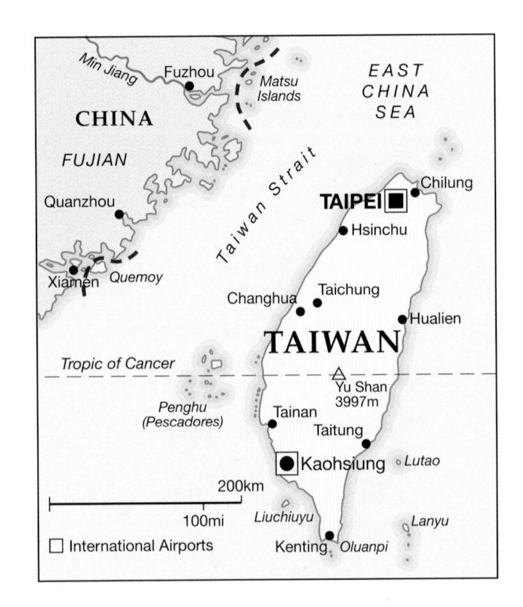

Facts and Figures

Capital: Taipei 2,720,000 (1992)
Area (sq km): 36,100
Population: 20,600,000 (1991)
Language: Mandarin, Chinese
Religion: Buddhist 30% Taoist 18% Christian 4%
Currency: New Taiwan Dollar = 100 cents
Annual Income per person: $8,815
Annual Trade per person: $7,395
Adult Literacy: 91%
Life Expectancy (F): 77
Life Expectancy (M): 72

TAJIKISTAN

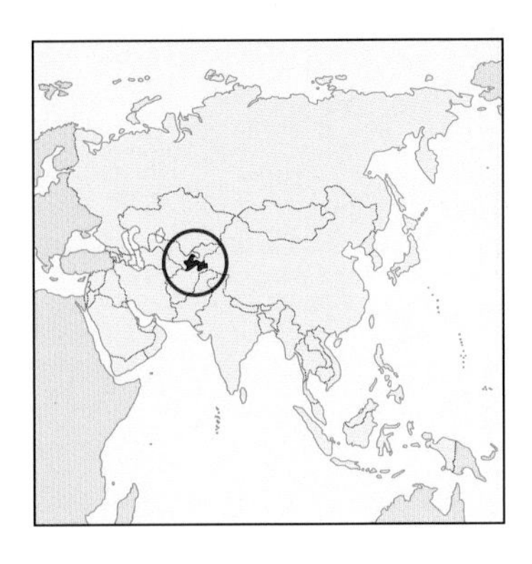

Facts and Figures

Capital: Dushanbe 592,000 (1991)
Area (sq km): 143,100
Population: 5,500,000(1993)
Language: Tajik, Uzbek, Russian
Religion: Sunni Muslim
Currency: Rouble = 100 kopecks
Annual Income per person: $1,050
Annual Trade per person: $400
Adult Literacy: Data not available
Life Expectancy (F): 72
Life Expectancy (M): 67

TURKMENISTAN

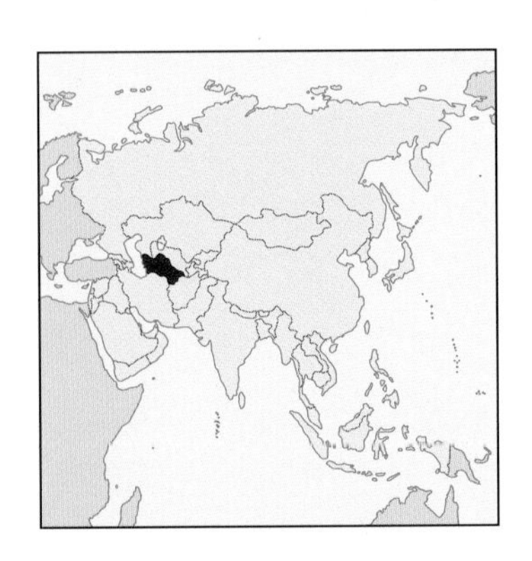

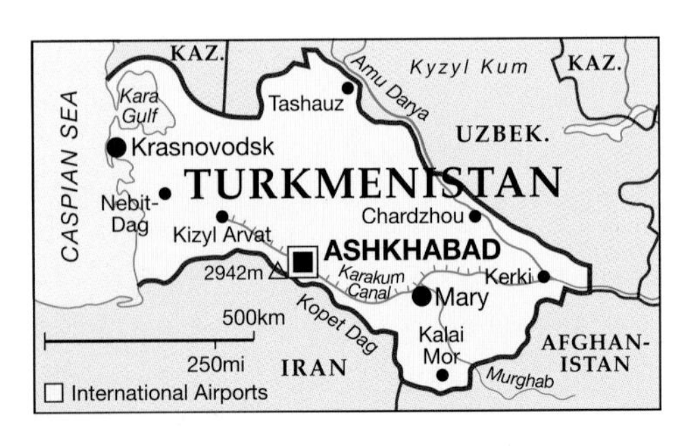

Facts and Figures

Capital: Ashkhabad 411,000 (1990)
Area (sq km): 488,100
Population: 3,800,000 (1992)
Language: Turkmen, Russian
Religion: Muslim 85%
Currency: Manat
Annual Income per person: $1,700
Annual Trade per person: $350
Adult Literacy: Data not available
Life Expectancy (F): 70
Life Expectancy (M): 63

THAILAND

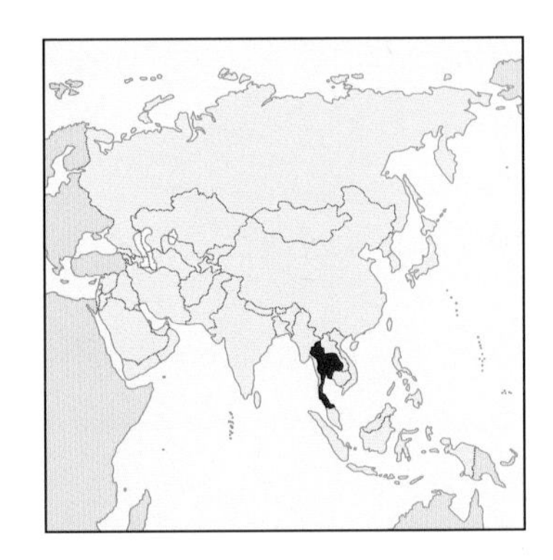

Facts and Figures

Capital: Bangkok 5,876,000 (1990)
Area (sq km): 513,115
Population: 57,800,000 (1993)
Language: Thai, Lao, Chinese, Malay
Religion: Buddhist 94% Muslim 4%
Currency: Baht = 100 satang
Annual Income per person: $1,580
Annual Trade per person: $1,152
Adult Literacy: 93%
Life Expectancy (F): 69
Life Expectancy (M): 65

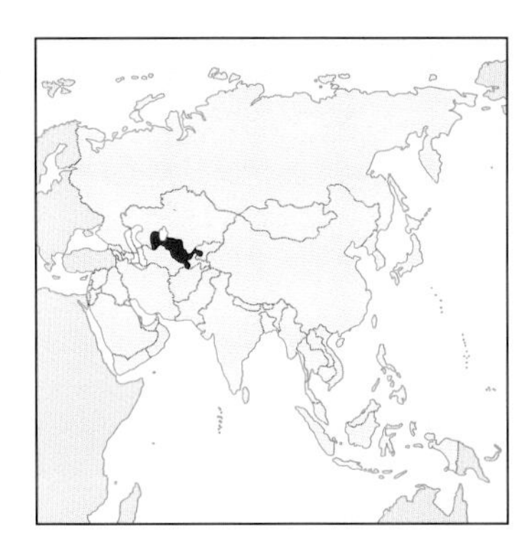

UZBEKISTAN

Facts and Figures

Capital: Tashkent 2,115,000 (1991)
Area (sq km): 447,400
Population: 21,200,000 (1992)
Language: Uzbek, Russian, Tajik
Religion: Sunni Muslim 75%
Currency: Rouble = 100 kopeks
Annual Income per person: $1,350
Annual Trade per person: $250
Adult Literacy: 93%
Life Expectancy (F): 73
Life Expectancy (M): 66

VIETNAM

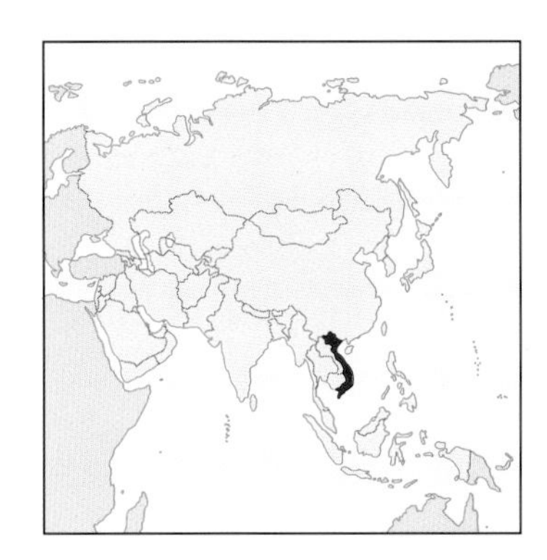

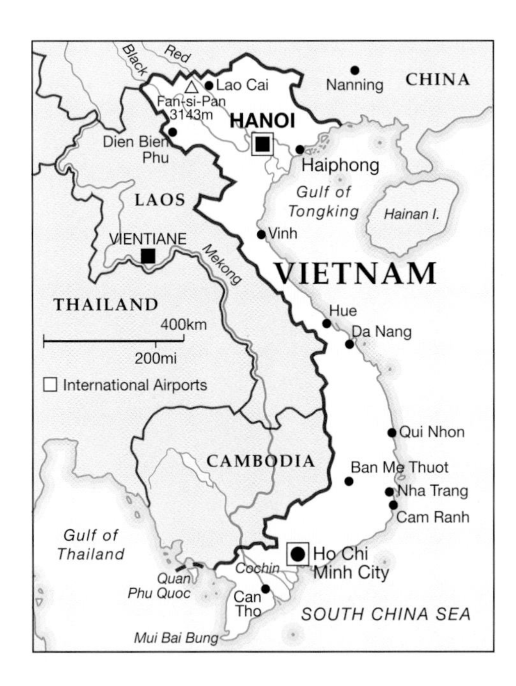

Facts and Figures

Capital: Hanoi 2,100,000 (1992)
Area (sq km): 329,570
Population: 69,300,000 (1992)
Language: Vietnamese
Religion: Buddhist 52% RC 7% Taoist
Currency: Dong = 10 hao
Annual Income per person: $300
Annual Trade per person: $60
Adult Literacy: 88%
Life Expectancy (F): 66
Life Expectancy (M): 62

UNITED ARAB EMIRATES

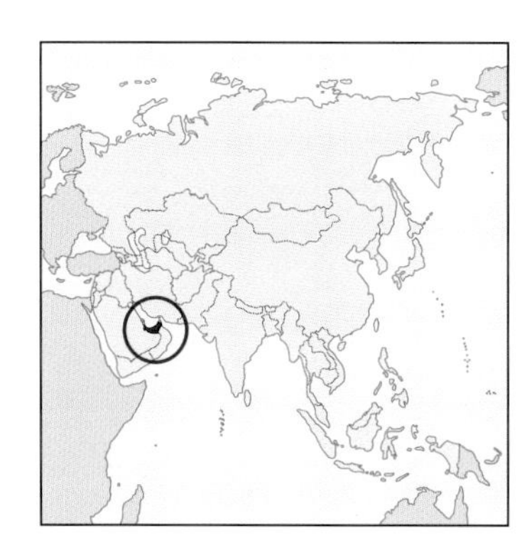

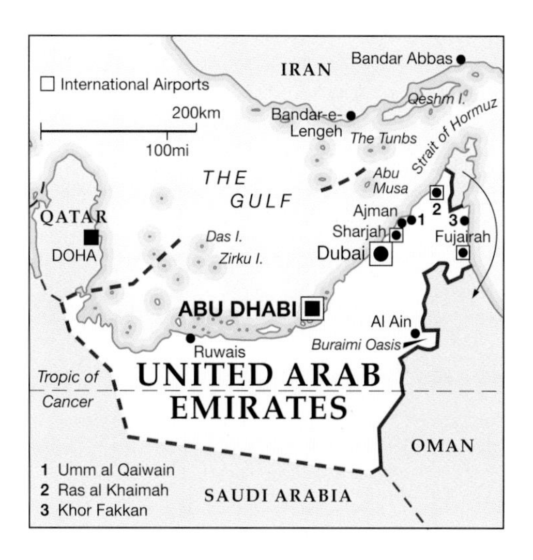

Facts and Figures

Capital: Abu Dhabi 806,000 (1992)
Area (sq km): 83,657
Population: 2,100,000 (1993)
Language: Arabic, English
Religion: Sunni Muslim 96%
Currency: Dirham = 100 fils
Annual Income per person: $19,870
Annual Trade per person: $19,745
Adult Literacy: 55%
Life Expectancy (F): 74
Life Expectancy (M): 70

YEMEN

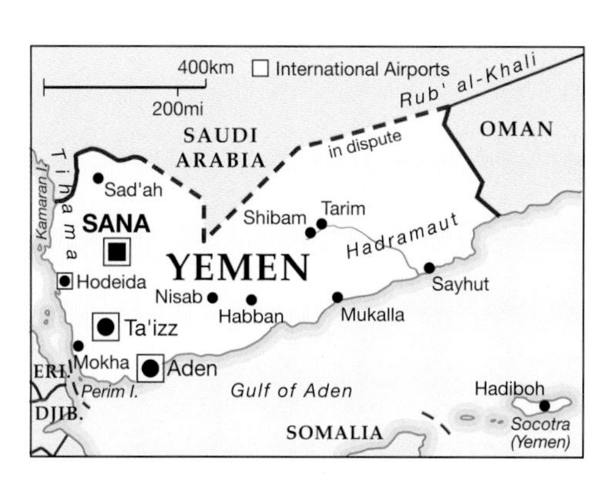

Facts and Figures

Capital: Sana 500,000 (1990)
Area (sq km): 531,000
Population: 13,000,000 (1993)
Language: Arabic
Religion: Sunni Muslim 45%
Shi'ite Muslim 41%
Currency: N. Yemeni Riyal and S. Yemeni Dinar
Annual Income per person: $540
Annual Trade per person: $141
Adult Literacy: 39%
Life Expectancy (F): 52
Life Expectancy (M): 52

OCEANIA

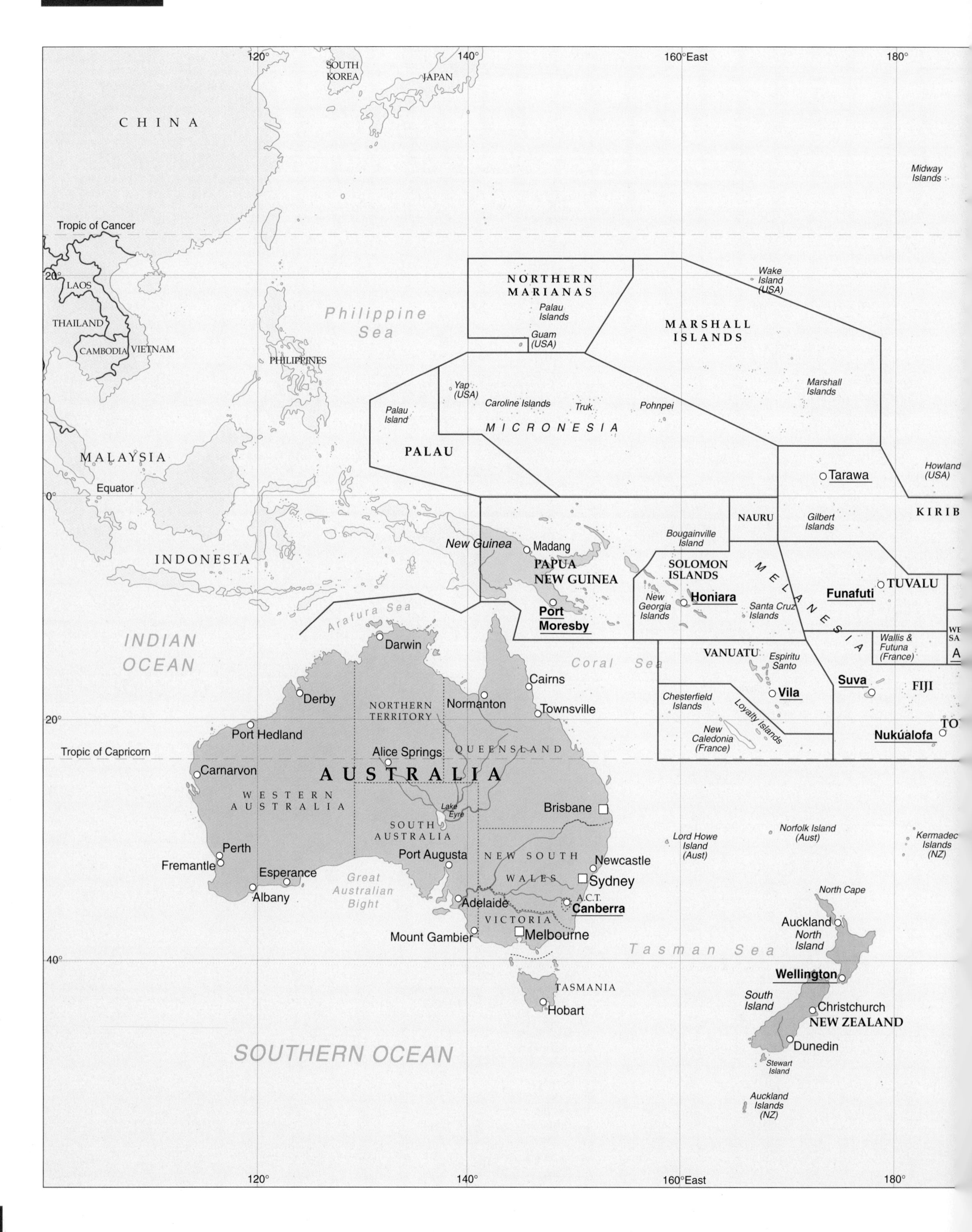

120°
140°
160°East
180°
SOUTH KOREA
JAPAN
CHINA
Midway Islands
Tropic of Cancer
20°
LAOS
THAILAND
CAMBODIA
VIETNAM
Philippine Sea
PHILIPPINES
NORTHERN MARIANAS
Palau Islands
Guam (USA)
Wake Island (USA)
MARSHALL ISLANDS
Marshall Islands
Yap (USA)
Palau Island
Caroline Islands
Truk
Pohnpei
MICRONESIA
PALAU
MALAYSIA
Equator
0°
Tarawa
Howland (USA)
NAURU
Gilbert Islands
KIRIB
Bougainville Island
New Guinea
Madang
PAPUA NEW GUINEA
INDONESIA
SOLOMON ISLANDS
MELANESIA
New Georgia Islands
Honiara
Santa Cruz Islands
Funafuti
TUVALU
Port Moresby
Arafura Sea
INDIAN OCEAN
Darwin
VANUATU
Espiritu Santo
Wallis & Futuna (France)
Coral Sea
Cairns
Suva
FIJI
Derby
NORTHERN TERRITORY
Normanton
Townsville
Chesterfield Islands
Vila
Loyalty Islands
New Caledonia (France)
Port Hedland
Nukúalofa
Tropic of Capricorn
Alice Springs
QUEENSLAND
Carnarvon
AUSTRALIA
WESTERN AUSTRALIA
Lake Eyre
Brisbane
SOUTH AUSTRALIA
Lord Howe Island (Aust)
Norfolk Island (Aust)
Kermadec Islands (NZ)
Perth
Fremantle
Port Augusta
NEW SOUTH WALES
Newcastle
Esperance
Sydney
Great Australian Bight
North Cape
Albany
Adelaide
A.C.T.
Canberra
VICTORIA
Auckland
North Island
Mount Gambier
Melbourne
Tasman Sea
40°
Wellington
TASMANIA
South Island
Christchurch
Hobart
NEW ZEALAND
Dunedin
SOUTHERN OCEAN
Stewart Island
Auckland Islands (NZ)

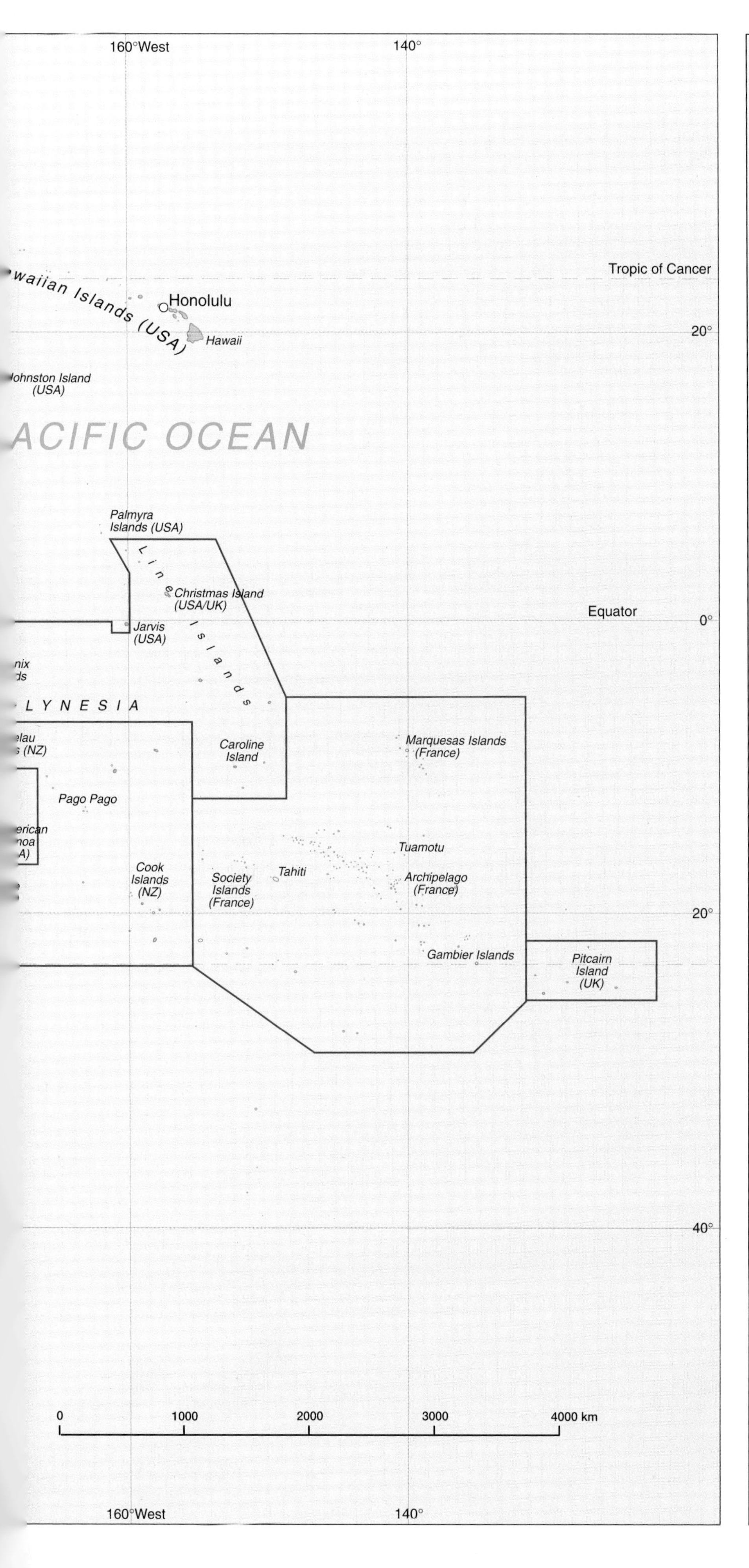
160°West
140°
Tropic of Cancer
Hawaiian Islands (USA)
Honolulu
Hawaii
20°
Johnston Island (USA)
PACIFIC OCEAN
Palmyra Islands (USA)
Line Islands
Christmas Island (USA/UK)
Equator
Jarvis (USA)
0°
POLYNESIA
Caroline Island
Marquesas Islands (France)
Pago Pago
Tuamotu
Cook Islands (NZ)
Society Islands (France)
Tahiti
Archipelago (France)
20°
Gambier Islands
Pitcairn Island (UK)
40°
0
1000
2000
3000
4000 km
160°West
140°

COUNTRY	PAGE
Australia	88
Fiji	88
Kiribati	88
New Zealand	89
Northern Marianas	89
Nauru	89
Marshall Islands	89
Papua New Guinea	90
Palau	90
Solomon Islands	90
Tonga	91
Tuvalu	88
Vanuatu	91
Western Samoa	91

AUSTRALIA

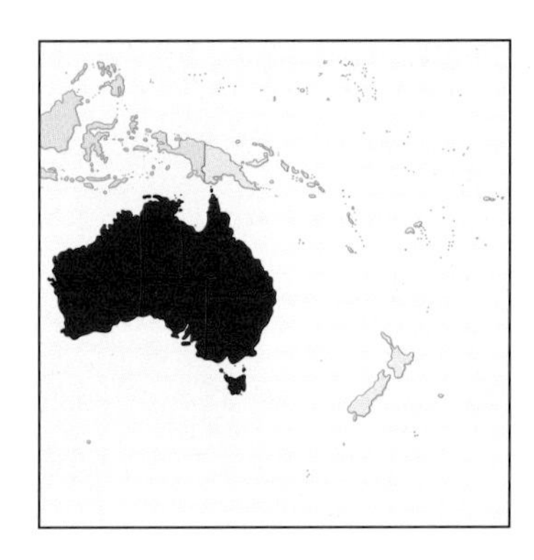

Facts and Figures

Capital: Canberra 278,000 (1991)
Area (sq km): 7,682,300
Population: 17,500,000 (1992)
Language: English
Religion: Catholic 26% Anglican 24% Other Christian 16%
Currency: Australian Dollar
Annual Income per person: $16,590
Annual Trade per person: $4,741
Adult Literacy: 99%
Life Expectancy (F): 80
Life Expectancy (M): 74

FIJI

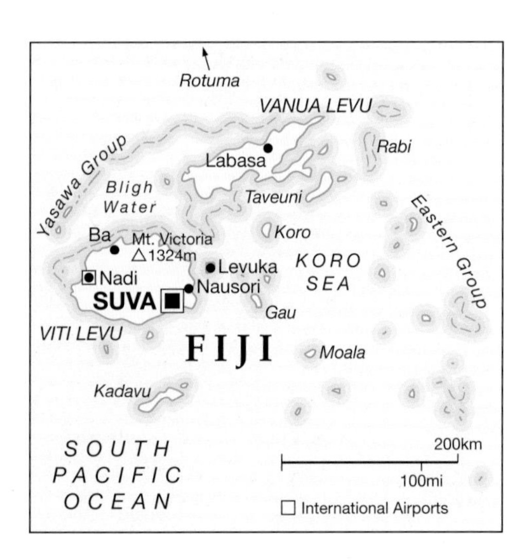

Facts and Figures

Capital: Suva 73,500 (1992)
Area (sq km): 18,333
Population: 758,275 (1993)
Language: English, Bauan, Hindustani
Religion: Christian 50% Hindu 36% Muslim 7%
Currency: Fiji dollar = 100 cents
Annual Income per person: $1,830
Annual Trade per person: $1,491
Adult Literacy: 87%
Life Expectancy (F): 68
Life Expectancy (M): 64

KIRIBATI & TUVALU

20km
10mi
TARAWA
Lagoon
Bonriki
Eita
Bairiki

London
20km
10mi
Paris
Bay of Wrecks
KIRITIMATI
(Christmas I.)

PACIFIC
Tabuaeran
Line Islands
Equator
Gilbert Is.
Phoenix Is.
KIRIBATI
Ellice Is.
SAMOAS
OCEAN
International Airports
2000km
1000mi

Nanumea Niutao
Nanumanga
TUVALU
Nui
Nukufetau
Vaitapu
FUNAFUTI
VAIAKU
200km
100mi
Nukulaelae
Niulakita

Facts and Figures

KIRIBATI

Capital: Bairiki (Tarawa) 25,200
Area (sq km): 720
Population: 72,300 (1991)
Language: English, Gilbertese
Religion: Roman Catholic 53% Protestant 39%
Currency: Australian Dollar
Annual Income per person: $750
Annual Trade per person: $350
Adult Literacy: Data not available
Life Expectancy (F): 58
Life Expectancy (M): 52

Facts and Figures

TUVALU

Capital: Funafuti 3,100 (1992)
Area (sq km): 24
Population: 10,090 (1991)
Language: Tuvaluan, English
Religion: Christian
Currency: Australian Dollar
Annual Income per person: $550
Annual Trade per person: $400
Adult Literacy: 95%
Life Expectancy (F): 63
Life Expectancy (M): 61

MARSHALL ISLANDS

Facts and Figures

Capital: Dalap-Uliga-Darrit 20,000 (1990)
Area (sq km): 181
Population: 45,600 (1990)
Language: Marshallese, English
Religion: Protestant
Currency: US Dollar
Annual Income per person: $1,500
Annual Trade per person: $700
Adult Literacy: 93%
Life Expectancy (F): 64
Life Expectancy (M): 61

NAURU

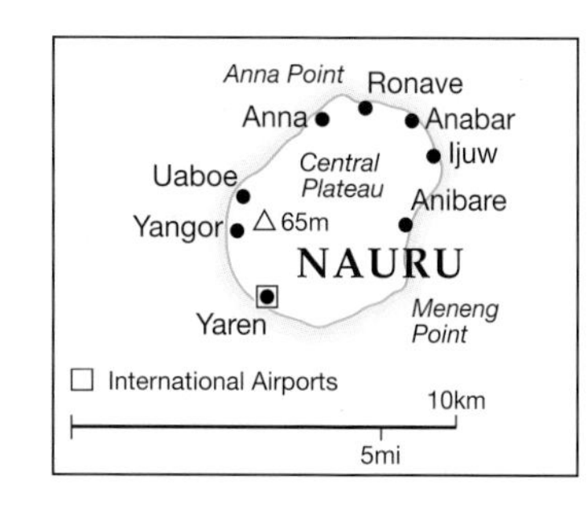

Facts and Figures

Capital: Yaren 7,500 (1990)
Area (sq km): 21
Population: 8,100 (1990)
Language: English, Nauruan
Religion: Roman Catholic, Protestant
Currency: Australian Dollar
Annual Income per person: 10,000
Annual Trade per person: $18,000
Adult Literacy: Data not available
Life Expectancy (F): 69
Life Expectancy (M): 64

NEW ZEALAND

Facts and Figures

Capital: Wellington 325,700 (1992)
Area (sq km): 270,530
Population: 3,495,000 (1993)
Language: English, Maori
Religion: Protestant 44% Roman Catholic 14% Other 7%
Currency: New Zealand Dollar
Annual Income per person: $12,140
Annual Trade per person: $5,615
Adult Literacy: 99%
Life Expectancy (F): 79
Life Expectancy (M): 73

NORTHERN MARIANAS

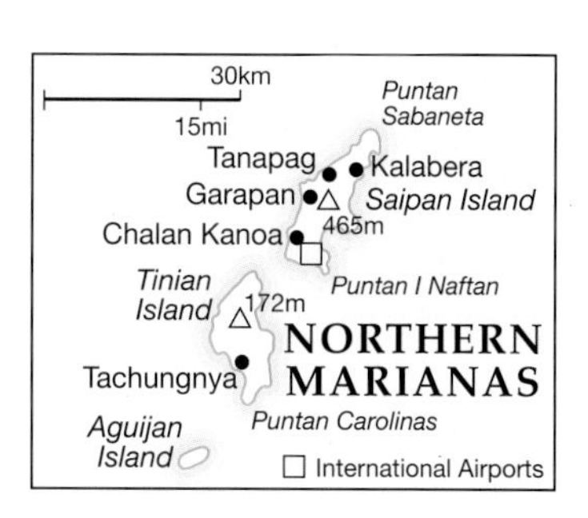

Facts and Figures

Capital: Saipan 38,900 (1990)
Area (sq km): 477
Population: 45.200 (1991)
Language: English, Chamorro
Religion: Roman Catholic
Currency: US Dollar
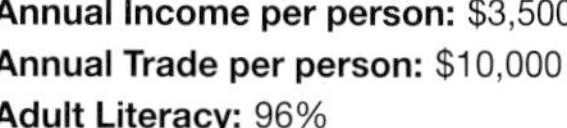
Annual Income per person: $3,500
Annual Trade per person: $10,000
Adult Literacy: 96%
Life Expectancy (F): 70
Life Expectancy (M): 65

PALAU

Facts and Figures

Capital: Koror 10,500 (1990)
Area (sq km): 488
Population: 15,200 (1990)
Language: Palaun, English
Religion: Roman Catholic
Currency: US Dollar
Annual Income per person: $2,550
Annual Trade per person: $1,922
Adult Literacy: Data not available
Life Expectancy (F): 74
Life Expectancy (M): 68

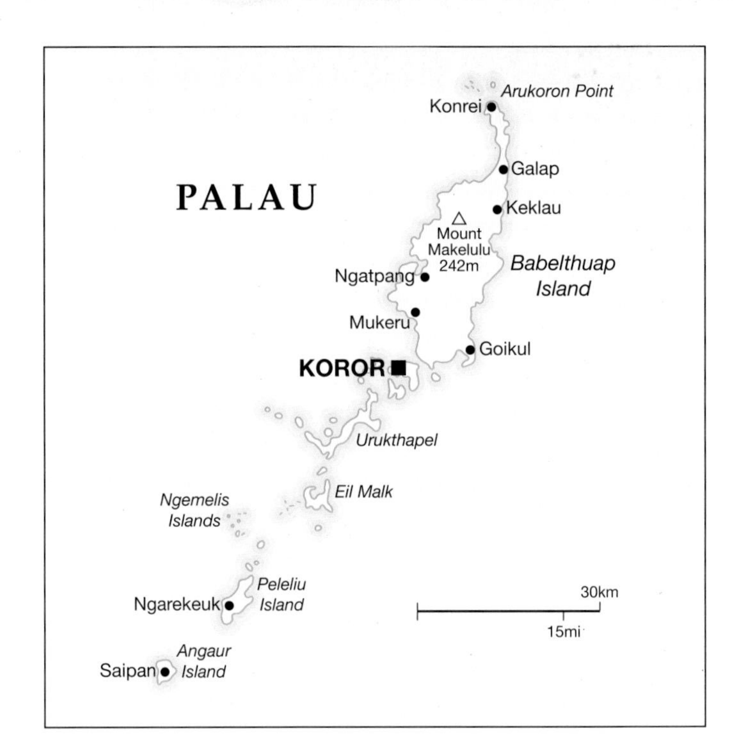

PAPUA NEW GUINEA

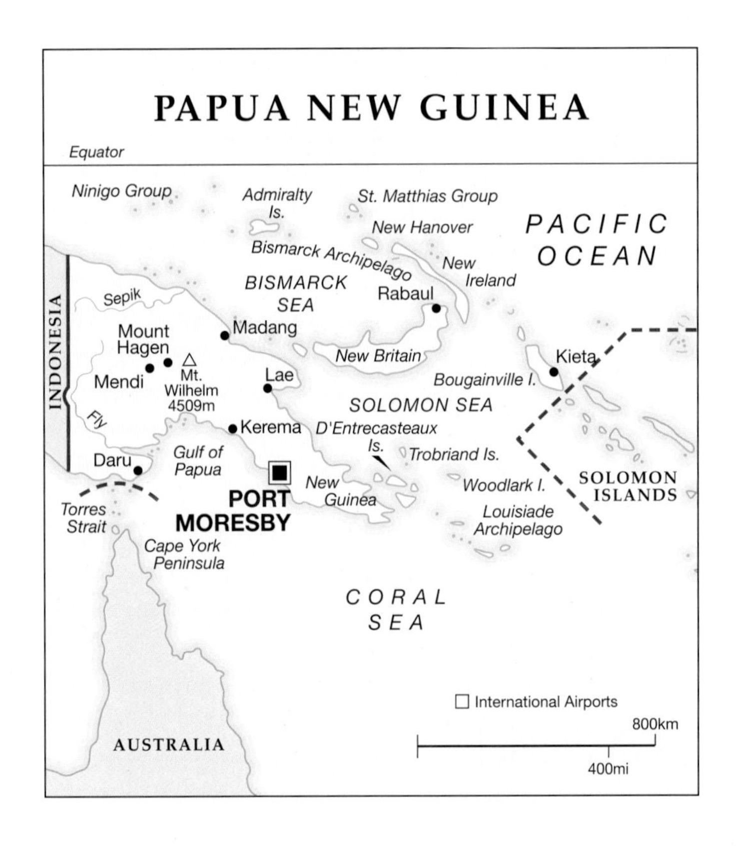

Facts and Figures

Capital: Port Moresby 193,300 (1990)
Area (sq km): 462,840
Population: 3,850,000 (1992)
Language: English, Aira Motu, Pidgin
Religion: Protestant 56% Roman Catholic 31% Traditional
Currency: Kina = 100 toea
Annual Income per person: $820
Annual Trade per person: $712
Adult Literacy: 52%
Life Expectancy (F): 57
Life Expectancy (M): 55

SOLOMON ISLANDS

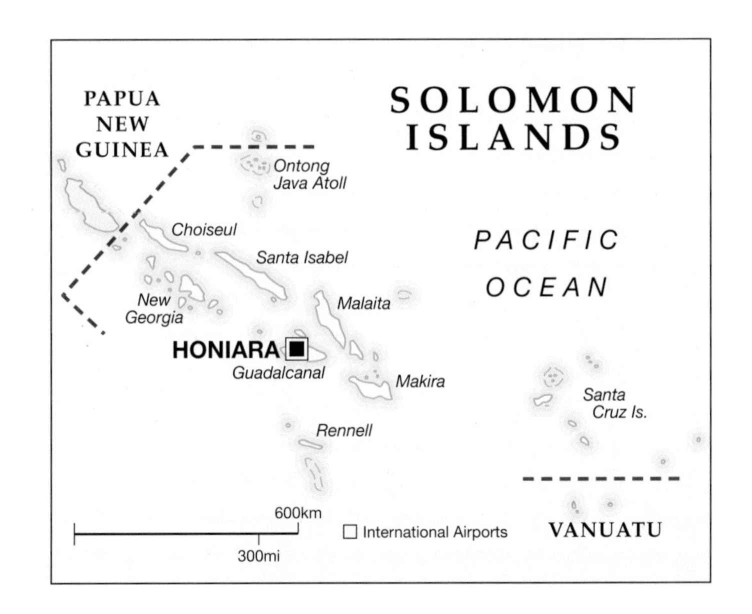

Facts and Figures

Capital: Honiara 33,800 (1989)
Area (sq km): 28,400
Population: 349,500 (1993)
Language: English, various Melanesian, Papuan and Polynesian languages
Religion: Protestant 75% Roman Catholic 19%
Currency: Solomon Is. Dollar = 100 cents
Annual Income per person: $560
Annual Trade per person: $506
Adult Literacy: 24%
Life Expectancy (F): 72
Life Expectancy (M): 67

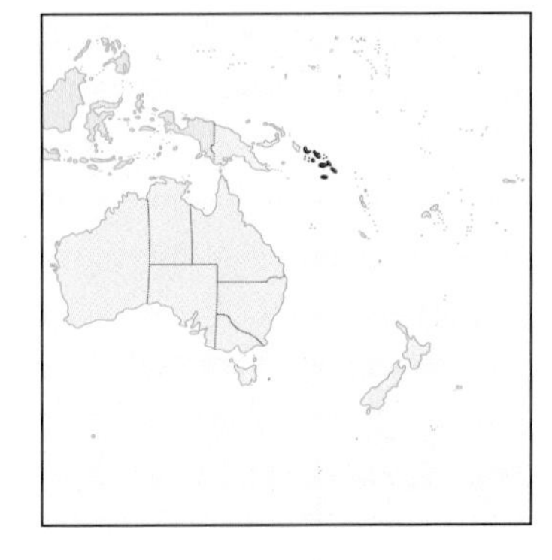

Facts and Figures

Capital: Nuku'alofa 29,000 (1986)
Area (sq km): 748
Population: 103,000 (1991)
Language: Tongan, English
Religion: Protestant 39%
Currency: Pa'anga = 100 seniti
Annual Income per person: $1,100
Annual Trade per person: $811
Adult Literacy: 99%
Life Expectancy (F): 70
Life Expectancy (M): 65

TONGA

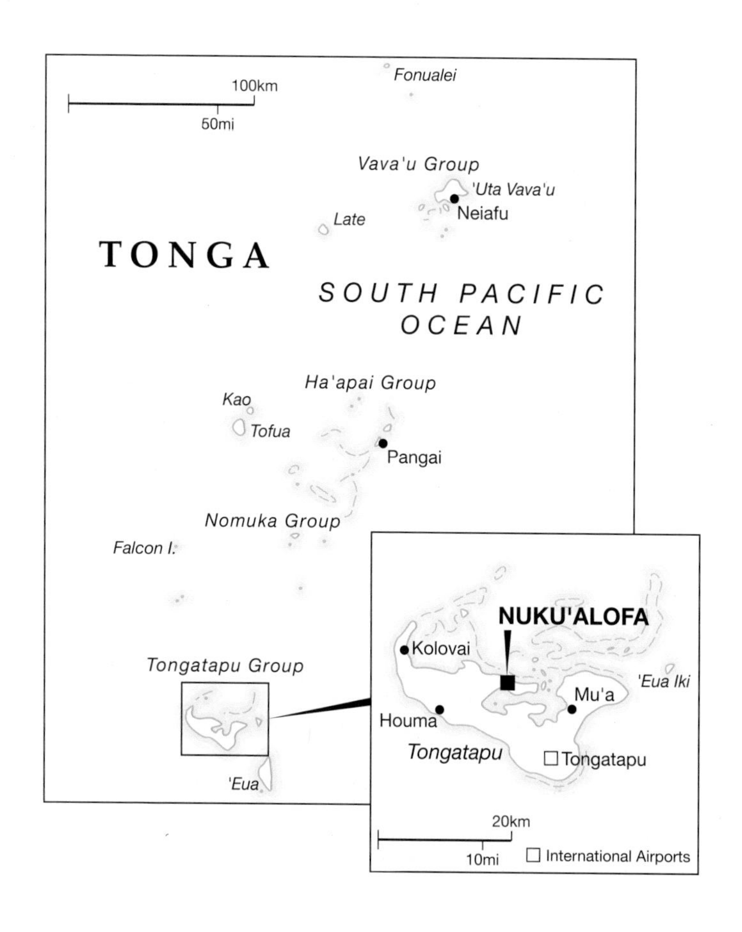

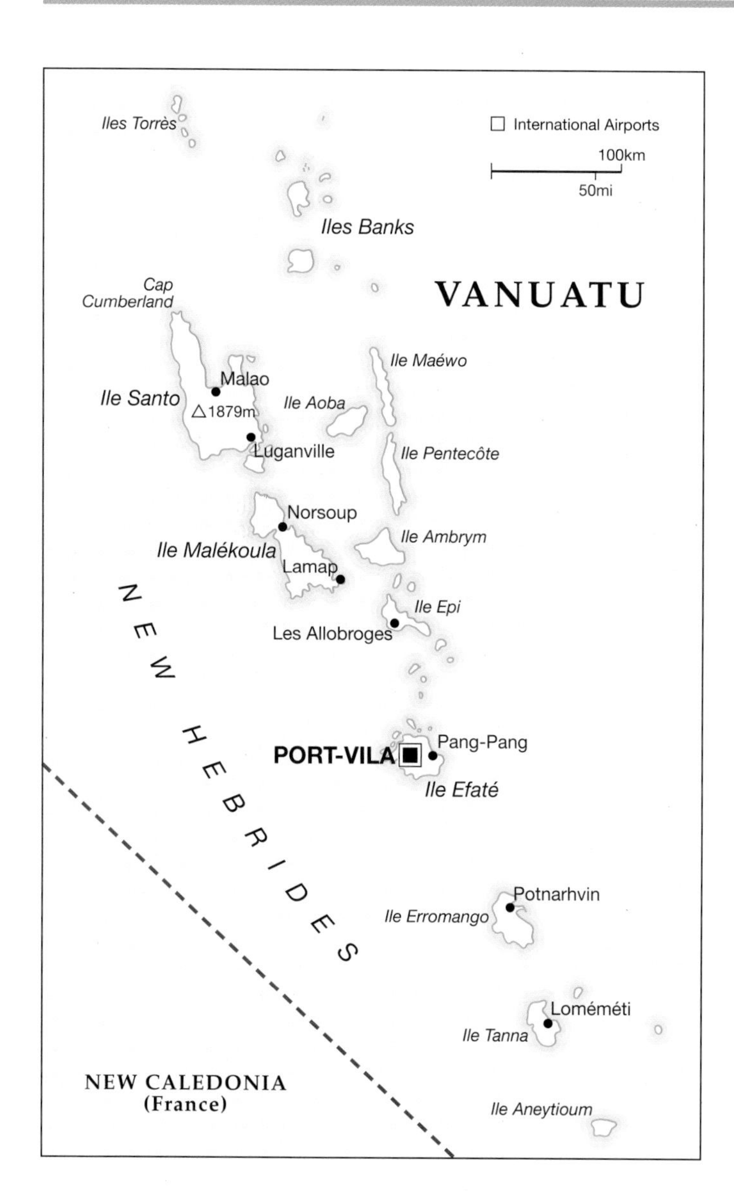

VANUATU

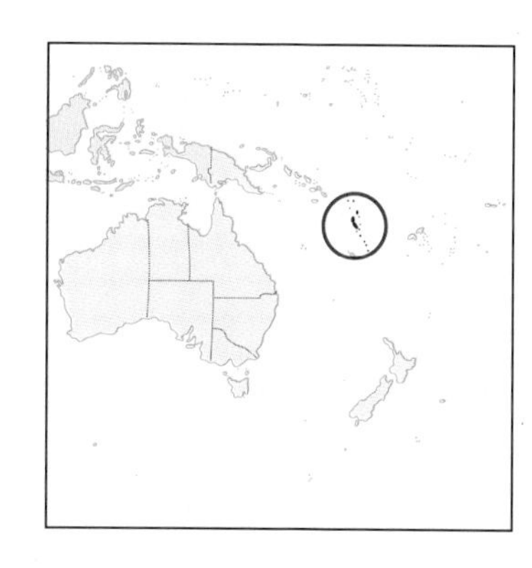

Facts and Figures

Capital: Port-Vila 19,400 (1989)
Area (sq km): 12,190
Population: 154,000 (1992)
Language: Bislama, English, French
Religion: Christian
Currency: Vatu = 100 centimes
Annual Income per person: $1,120
Annual Trade per person: $644
Adult Literacy: 67%
Life Expectancy (F): 72
Life Expectancy (M): 67

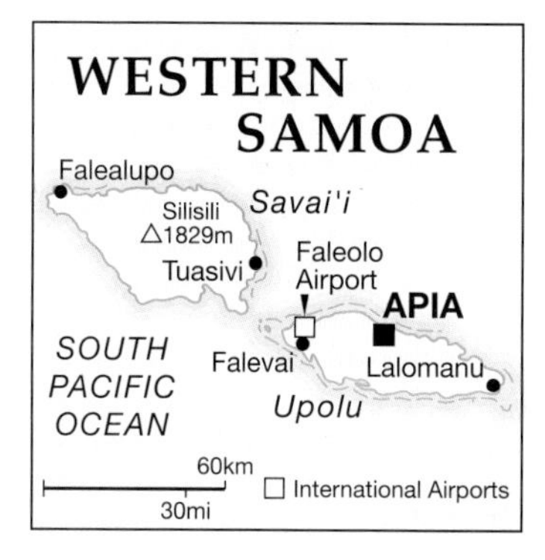

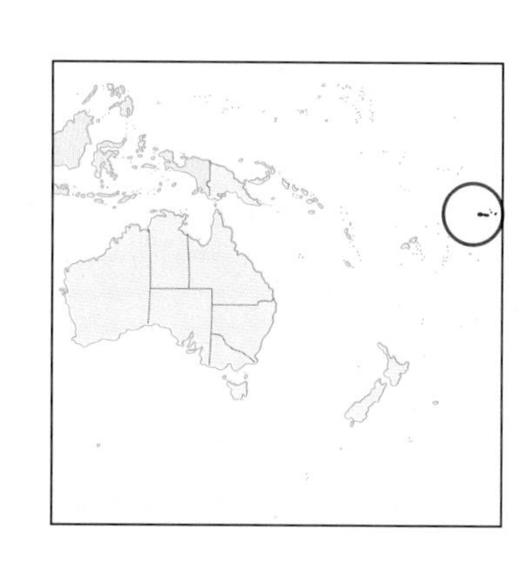

WESTERN SAMOA

Facts and Figures

Capital: Apia 32,200 (1986)
Area (sq km): 2,830
Population: 168,000 (1990)
Language: Samoan, English
Religion: Protestant 62%
Roman Catholic 22%
Currency: Tala = 100 sene
Annual Income per person: $930
Annual Trade per person: $629
Adult Literacy: 92%
Life Expectancy (F): 69
Life Expectancy (M): 64

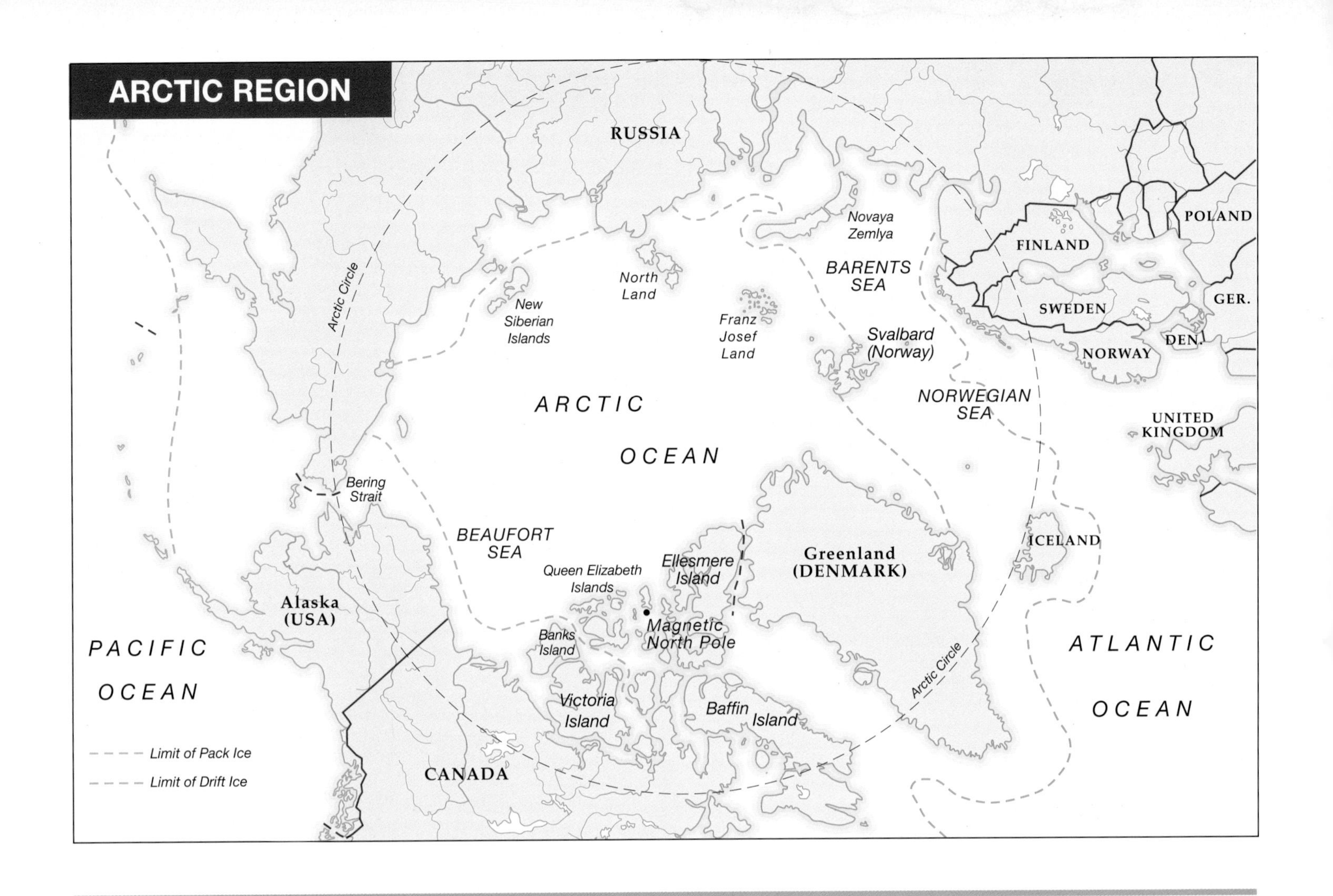
ARCTIC REGION
RUSSIA
Novaya Zemlya
FINLAND
POLAND
BARENTS SEA
North Land
New Siberian Islands
Franz Josef Land
Svalbard (Norway)
SWEDEN
GER.
NORWAY
DEN.
Arctic Circle
ARCTIC OCEAN
NORWEGIAN SEA
UNITED KINGDOM
Bering Strait
BEAUFORT SEA
ICELAND
Queen Elizabeth Islands
Ellesmere Island
Greenland (DENMARK)
Alaska (USA)
Magnetic North Pole
Banks Island
PACIFIC OCEAN
ATLANTIC OCEAN
Victoria Island
Baffin Island
Arctic Circle
Limit of Pack Ice
Limit of Drift Ice
CANADA

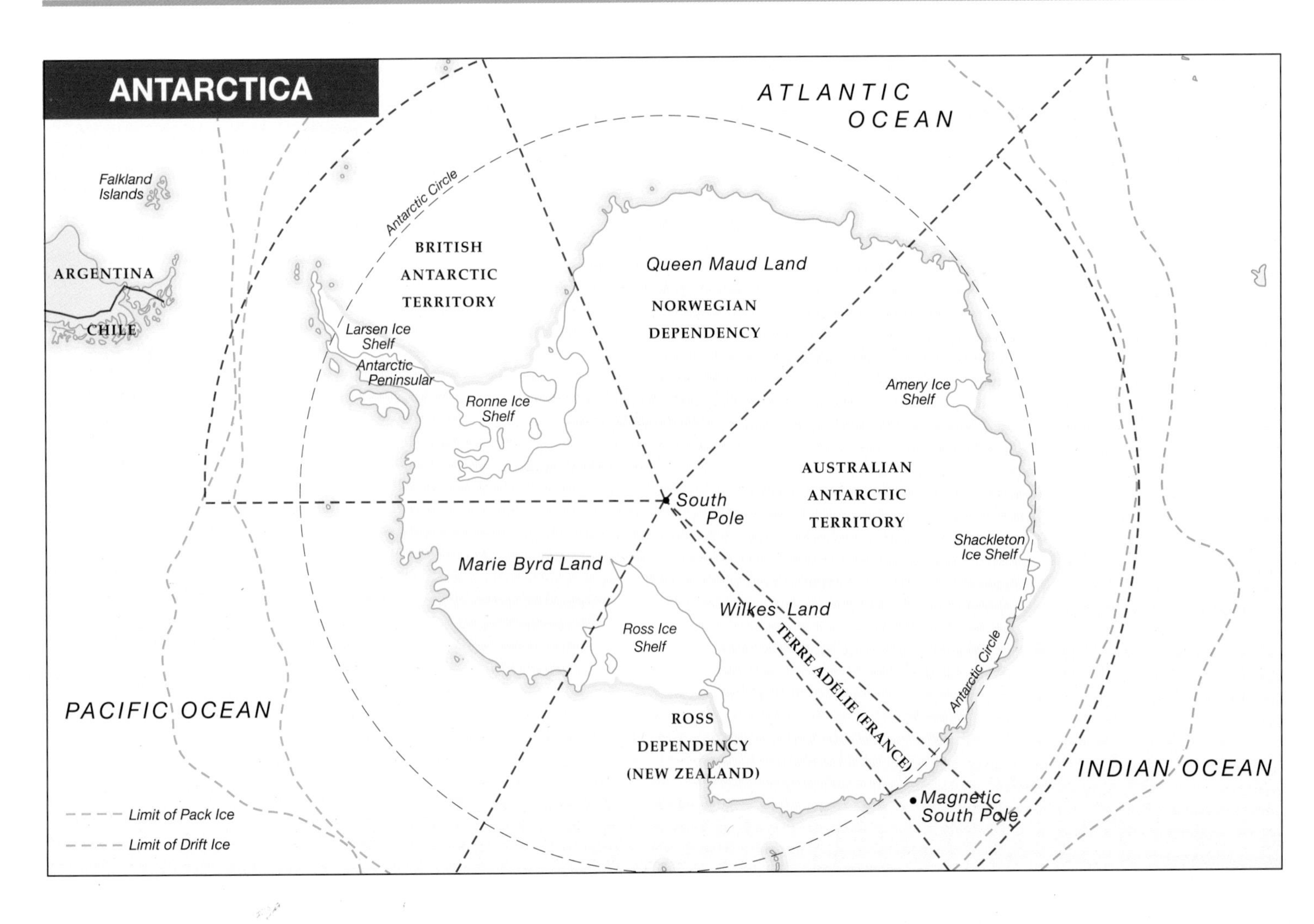
ANTARCTICA
ATLANTIC OCEAN
Falkland Islands
Antarctic Circle
BRITISH ANTARCTIC TERRITORY
Queen Maud Land
ARGENTINA
NORWEGIAN DEPENDENCY
CHILE
Larsen Ice Shelf
Antarctic Peninsular
Amery Ice Shelf
Ronne Ice Shelf
AUSTRALIAN ANTARCTIC TERRITORY
South Pole
Shackleton Ice Shelf
Marie Byrd Land
Wilkes Land
Ross Ice Shelf
TERRE ADÉLIE (FRANCE)
Antarctic Circle
PACIFIC OCEAN
ROSS DEPENDENCY (NEW ZEALAND)
INDIAN OCEAN
Magnetic South Pole
Limit of Pack Ice
Limit of Drift Ice

INDEX

A

Aachen Germany 21
Abadan Iran 74
Abéché Chad 53
Abeokuta Nigeria 62
Aberdeen United Kingdom 31
Aberystwyth United Kingdom 31
Abidjan Côte D'Ivoire 54
Abomey Benin 51
Abu Dhabi United Arab Emirates 85
Abu Kamal Syria 83
Abuja Nigeria 62
Acapulco Mexico 39
Accra Ghana 56
Acre Israel 75
Ada Ghana 56
Adana Turkey 29
Addis Ababa Ethiopia 55
Adelaide Australia 88
Aden Yemen 85
Adrar Algeria 50
Agadez Niger 62
Agadir Morocco 61
Agdam Azerbaijan 70
Agra India 74
Ahmadabad India 74
Ahvaz Iran 74
Ajman United Arab Emirates 85
Akmola Kazakhstan 77
Akranes Iceland 22
Akrotiri Cyprus 19
Aksum Ethiopia 55
Aktau Kazakhstan 77
Aktogay Kazakhstan 77
Aktyubinsk Kazakhstan 77
Akureyri Iceland 22
Al Ahmadi Kuwait 77
Al Ain United Arab Emirates 85
Al Basrah Iraq 75
Al Bayda Libya 58
Al Hasakah Syria 83
Al Jafr Jordan 76
Al Jahrah Kuwait 77
Al Khawr Qatar 81
Al Kufrah Libya 58
Al Kut Iraq 75
Al Wakrah Qatar 81
Alajuela Costa Rica 35
Alanya Turkey 29
Alaverdi Armenia 70
Ålborg Denmark 19
Albreda Gambia 56
Albuquerque United States 41
Ålesund Norway 26
Alexander Bay South Africa 64
Alexandria Egypt 55
Algiers Algeria 50
Ali-Sabieh Djibouti 54
Alicante Spain 28
Alice Springs Australia 88
All Saints Antigua & Barbuda 34
Alma-Ata Kazakhstan 77
Almería Spain 28
Alta Norway 26
Altyus Lithuania 24
Am Timan Chad 53
Ambohimanga Madagascar 59
Amiens France 20
Amman Jordan 76
Amritsar India 74
Amsterdam Netherlands 26
An Han Uzbekistan 85
An Najaf Iraq 75
Anabar Nauru 89
Anadyr Russia 82
Anchorage United States 41
Andermatt Switzerland 29
Andorra-la-Vella Andorra 16
Andros Town Bahamas 34
Aného Togo 65
Angekhakot Armenia 70
Angmagssalik Greenland 37
Angoche Mozambique 61
Angra do Heroismo Azores 50
Anibare Nauru 89
Ankara Turkey 29
Anna Nauru 89
Annaba Algeria 50
Antakya Turkey 29
Antalya Turkey 29
Antananarivo Madagascar 59
Antigua Guatemala 38
Antofagasta Chile 45
Antseranana Madagascar 59
Antwerp Belgium 17
Anuradhapura Sri Lanka 83
Aomori Japan 75
Apia Western Samoa 91
Apoteri Guyana 46
Aqaba Jordan 76
Arad Romania 27
Aralsk Kazakhstan 77
Ararat Armenia 70
Arbil Iraq 75
Arcalis Andorra 16
Arecibo Puerto Rico 40
Arequipa Peru 46
Argyle St Vincent 40
Århus Denmark 19
Arica Chile 45
Arima Trinidad & Tobago 41
Arkhangelsk Russia 82
Arlon Belgium 17
Armagh United Kingdom 31
Arnhem Netherlands 26
Arrecife Canary Islands 52
Arua Uganda 66
Arusha Tanzania 65
Aseb Eritrea 55
Ashkhabad Turkmenistan 84
Asmara Eritrea 55
Asunción Paraguay 46
Aswan Egypt 55
Atakpamé Togo 65
Atar Mauritania 60
Atbara Sudan 64
Atbasar Kazakhstan 77
Athens Greece 22
Athimatha Maldives 79
Atlanta United States 41
Atyrau Kazakhstan 77
Auckland New Zealand 89
Aveiro Portugal 27
Awali Bahrain 71
Awash Ethiopia 55
Ayorou Niger 62
Ayr United Kingdom 31
Azraq Jordan 76

B

Ba Fiji 88
Baalbek Lebanon 78
Bacolod Philippines 81
Badulla Sri Lanka 83
Bafatá Guinea Bissau 57
Baghdad Iraq 75
Baguio Philippines 81
Bahía Blanca Argentina 44
Bahlah Oman 80
Baikonur Kazakhstan 77
Bairiki Kiribati 88
Bakau Gambia 56
Bakel Senegal 63
Bakhtaran Iran 74
Baku Azerbaijan 70
Balikpapan Indonesia 74
Balkh Afghanistan 70
Balkhash Kazakhstan 77
Ballymena United Kingdom 31
Baltimore United States 41
Balzers Liechtenstein 24
Bamako Mali 60
Bambari Central African Republic 53
Bamiyan Afghanistan 70
Ban Hat Yai Thailand 84
Ban Houei Sai Laos 78
Ban Kantang Thailand 84
Ban Me Thuot Vietnam 85
Bandar Abbas Iran 74
Bandar Seri Begawan Brunei 72
Bandiagara Mali 60
Bangalore India 74
Bangar Brunei 72
Bangassou Central African Republic 53
Bangkok Thailand 84
Bangor United Kingdom 31
Bangui Central African Republic 53
Banja Luka Bosnia-Herzegovina 18
Banjul Gambia 56
Banská Bystrica Slovak Republic 27
Barahona Dominican Republic 37
Baranovichi Belarus 17
Barcelona Spain 28
Barcelona Venezuela 47
Bari Italy 23
Barmouth United Kingdom 31
Barquisimeto Venezuela 47
Barranquilla Colombia 45
Barrouallie St Vincent 40
Bartica Guyana 46
Basel Switzerland 29
Bassar Togo 65
Basse Santa Su Gambia 56
Basseterre St Christopher/Nevis 40
Bata Equatorial Guinea 55
Batavia Surinam 47
Bath United Kingdom 30
Bath St Christopher/Nevis 40
Batouri Cameroon 52
Battambang Cambodia 72
Batticaloa Sri Lanka 83
Batumi Georgia 72
Bauchi Nigeria 62
Bayamón Puerto Rico 40
Bayir Jordan 76
Bayonne France 20
Béchar Algeria 50
Beersheba Israel 75
Beijing China 73
Beira Mozambique 61
Beirut Lebanon 78
Beitbridge Zimbabwe 67
Beja Portugal 27
Bel'tsy Moldova 25
Beledweyne Somalia 64
Belém Brazil 44
Belfast United Kingdom 31
Belgrade Yugoslavia 29
Belize City Belize 35
Belleplaine Barbados 34
Belmopan Belize 35
Belo Horizonte Brazil 44
Benghazi Libya 58
Benguela Angola 50
Benin City Nigeria 62
Berat Albania 16
Berbera Somalia 64
Berbérati Central African Republic 53
Berekua Dominica 36
Bergen Norway 26
Bergen op Zoom Netherlands 26
Berlin Germany 21
Bern Switzerland 29
Bétaré Oya Cameroon 52
Beynau Kazakhstan 77
Bhadgaon Nepal 80
Biabou St Vincent 40
Bialystók Poland 26
Bihac Bosnia-Herzegovina 18
Bijeljina Bosnia-Herzegovina 18
Bilbao Spain 28
Bintulu Malaysia 78
Bir Moghrein Mauritania 60
Birao Central African Republic 53
Birjand Iran 74
Birkirkara Malta 25
Birmingham United Kingdom 30
Bishkek Kyrgyzstan 77
Bissau Guinea Bissau 57
Bitola Macedonia 25
Biumba Rwanda 62
Bizerta Tunisia 66
Blackmans Barbados 34
Blantyre Malawi 59
Bled Slovenia 28
Blida Algeria 50
Bloemfontein South Africa 64
Bluefields Nicaragua 39
Bo Sierra Leone 63
Boali Central African Republic 53
Bobo Dioulasso Burkina Faso 51
Bocaranga Central African Republic 53
Bodø Norway 26
Bogotá Colombia 45
Bogra Bangladesh 71
Boké Guinea 57
Bolama Guinea Bissau 57
Bolgatanga Ghana 56
Bologna Italy 23
Bombay India 74
Bondo Zaïre 66
Bongor Chad 53
Bonn Germany 21
Bonriki Kiribati 88
Bonthe Sierra Leone 63
Bordeaux France 20
Borzhomi Georgia 72
Bossango Central African Republic 53
Boston United States 41
Bouaké Côte D'Ivoire 54
Bouar Central African Republic 53
Boulogne France 20
Bourke Australia 88
Boutilimit Mauritania 60
Braga Portugal 27
Bragança Portugal 27
Brasília Brazil 44
Brasov Romania 27
Bratislava Slovak Republic 27
Brazzaville Congo 54
Brecon United Kingdom 31
Bregenz Austria 16
Bremen Germany 21
Brest Belarus 17
Brezice Slovenia 28
Bridgetown Barbados 34
Brikama Gambia 56
Brindisi Italy 23
Brisbane Australia 88
Bristol United Kingdom 30
Brno Czech Republic 19
Broken Hill Australia 88
Broome Australia 88
Bruges Belgium 17
Brussels Belgium 17
Bubanza Burundi 52
Bucaramanga Colombia 45
Buchanan Liberia 58
Bucharest Romania 27
Budapest Hungary 22
Buenaventura Colombia 45
Buenos Aires Argentina 44
Builth Wells United Kingdom 31
Bujumbura Burundi 52
Bukavu Zaïre 66
Bukhara Uzbekistan 85
Bukit Panjang Singapore 83
Bulawayo Zimbabwe 67
Bura Kenya 57
Buraimi Oman 80
Burao Somalia 64
Burgas Bulgaria 18
Burgos Spain 28
Bursa Turkey 29
Bururi Burundi 52
Bushehr Iran 74
Butare Rwanda 62
Byblos Lebanon 78
Bydgoszcz Poland 26

C

Cabinda Angola 50
Cádiz Spain 28
Caernarfon United Kingdom 31
Cagayan de Oro Philippines 81
Cagliari Italy 23
Caguas Puerto Rico 40
Cairns Australia 88
Cairo Egypt 55
Cajamarca Peru 46
Calais France 20
Calcutta India 74
Calgary Canada 36
Cali Colombia 45
Callao Peru 46
Calliaqua St Vincent 40
Caluula Somalia 64
Cam Ranh Vietnam 85
Camagüey Cuba 36
Cambridge United Kingdom 30
Campo Grande Brazil 44
Can Tho Vietnam 85
Canaries St Lucia 40
Canberra Australia 88
Canchungo Guinea Bissau 57
Canillo Andorra 16
Cap Haïtien Haiti 38
Cape Coast Ghana 56
Cape Town South Africa 64
Caracas Venezuela 47
Cardiff United Kingdom 31
Cardigan United Kingdom 31
Cartagena Spain 28
Cartagena Colombia 45
Cartago Costa Rica 35
Casablanca Morocco 61
Castellón de la Plana Spain 28
Castelo Branco Portugal 27
Castle Bruce Dominica 36
Castries St Lucia 40
Catania Italy 23
Catio Guinea Bissau 57
Cayon St Christopher/Nevis 40
Cebu Philippines 81
Celje Slovenia 28
Ceské Budejovice Czech Republic 19
Ceuta Spain 28
Chalan Kanoa Northern Marianas 89
Chanea Greece 22
Changhua Taiwan 84
Changsha China 73
Chardzhou Turkmenistan 84
Charleroi Belgium 17
Charleston United States 41
Charlestown St Christopher/Nevis 40
Charleville Australia 88
Charlotte United States 41
Charlottetown Canada 36
Charlotteville Trinidad & Tobago 41
Chartres France 20
Chateaubelair St Vincent 40
Cheb Czech Republic 19
Cheju South Korea 76
Chelkar Kazakhstan 77
Chemnitz Germany 21
Chengde China 73
Chengdu China 73
Cherbourg France 20
Chernobyl Ukraine 29
Chiang Mai Thailand 84
Chiang Rai Thailand 84
Chibuto Mozambique 61
Chicago United States 41
Chiclayo Peru 46
Chihuahua Mexico 39
Chilumba Malawi 59
Chilung Taiwan 84
Chimkent Kazakhstan 77
Chimoio Mozambique 61
Chinandega Nicaragua 39
Chinde Mozambique 61
Chinguetti Mauritania 60
Chinhoyi Zimbabwe 67
Chipata Zambia 67
Chisinau Moldova 25
Chita Russia 82
Chittagong Bangladesh 71
Choiseul St Lucia 40
Choluteca Honduras 38
Chongjin North Korea 76
Chongqing China 73
Chonju South Korea 76
Choybalsan Mongolia 79
Christchurch New Zealand 89
Chu Kazakhstan 77
Chungju South Korea 76
Cienfuegos Cuba 36
Cincinnati United States 41
Ciudad Bolívar Venezuela 47
Ciudad Guayana Venezuela 47
Ciudad Real Spain 28
Clervaux Luxembourg 24
Clifton St Vincent 40
Cluj-Napoca Romania 27
Cobija Bolivia 44
Coca Ecuador 45
Cochabamba Bolivia 44
Cocoa Point Antigua & Barbuda 34
Codrington Antigua & Barbuda 34
Coimbra Portugal 27
ColeraineUnited Kingdom (N. Ireland) 31
Colombo Sri Lanka 83
Colón Panama 39
Colonarie St Vincent 40
Colwyn Bay United Kingdom (Wales) 31
Comilla Bangladesh 71
Comodoro Rivadavia Argentina 44
Conakry Guinea 57
Concepción Chile 45
Concepción Paraguay 46
Constanta Romania 27
Constantine Algeria 50
Copenhagen Denmark 19
Córdoba Spain 28
Córdoba Argentina 44
Corfu Greece 22
Corinto Nicaragua 39
Cork Ireland 23
Coro Venezuela 47
Corozal Belize 35
Cotonou Benin 51
Coulihaut Dominica 36
Coventry United Kingdom (England) 30
Covilhã Portugal 27
Cox's Bazar Bangladesh 71
Craiova Romania 27
Cristóbal Panama 39
Cruzeiro do Sul Brazil 44
Cuenca Ecuador 45
Cujda Morocco 61
Curepipe Mauritius 60
Cuttack India 74
Cyangugu Rwanda 62

D

Da Nang Vietnam 85
Daddato Djibouti 54
Dakar Senegal 63
Dakhla Western Sahara 61
Dalaman Turkey 29
Dalap-Uliga-Darrit Marshall Islands 89
Dalian China 73
Dallas United States 41
Daloa Côte D'Ivoire 54
Damman Saudi Arabia 82
Dangriga Belize 35
Dar es Salaam Tanzania 65
Dara Syria 83
Darhan Mongolia 79
Daru Papua New Guinea 90
Darwin Australia 88
Daugavpils Latvia 24
Davao Philippines 81
David Panama 39
Davos Switzerland 29
De Aar South Africa 64
Debar Macedonia 25
Debrecen Hungary 22
Delcovo Macedonia 25
Delgado El Salvador 37
Den Helder Netherlands 26
Dennery St Lucia 40
Denver United States 41
Derry United Kingdom 31
Detroit United States 41
Dhahran Saudi Arabia 82
Dhaka Bangladesh 71
Dhankuta Nepal 80
Dhekelia Cyprus 19
Dhlggiri Maldives 79
Dibrugarh India 74
Diekirch Luxembourg 24
Dieppe France 20
Dieppe Bay St Christopher/Nevis 40
Diffa Niger 62
Differdange Luxembourg 24
Dijon France 20
Dikhil Djibouti 54
Dilizhan Armenia 70
Dinajpur Bangladesh 71
Dinant Belgium 17
Diourbel Senegal 63
Dire Dawa Ethiopia 55
Djambala Congo 54
Djanet Algeria 50
Djenne Mali 60
Djibouti Djibouti 54
Djougou Benin 51
Dnepropetrovsk Ukraine 29
Dodoma Tanzania 65
Doha Qatar 81
Donegal Ireland 23
Donetsk Ukraine 29
Dongola Sudan 64
Dordrecht Netherlands 26
Dortmund Germany 21
Dosso Niger 62
Douala Cameroon 52
Dover United Kingdom 30
Dresden Germany 21
Drogheda Ireland 23
Druzhba Kazakhstan 77
Dubai United Arab Emirates 85
Dublin Ireland 23
Dubossary Moldova 25
Dubrovnik Croatia 18
Dudelange Luxembourg 24
Duisburg Germany 21
Dukhan Qatar 81
Dulcina Antigua & Barbuda 34
Dumfries United Kingdom 31
Dun Laoghaire Ireland 23
Dundalk Ireland 23
Dundas Greenland 37
Dundee United Kingdom 31
Dunedin New Zealand 89
Dunkwa Ghana 56
Durazno Uruguay 47
Durban South Africa 64
Durham United Kingdom 30
Durrës Albania 16
Dushanbe Tajikistan 84
Düsseldorf Germany 21
Dzhalal-Abad Kyrgyzstan 77
Dzhambul Kazakhstan 77
Dzhezkazgan Kazakhstan 77
Dzhusaly Kazakhstan 77

E

East London South Africa 64
Ebebiyin Equatorial Guinea 55
Echternach Luxembourg 24
Ed Eritrea 55
Edinburgh United Kingdom 31
Edirne Turkey 29
Edmonton Canada 36
Eilat Israel 75
Eindhoven Netherlands 26
Eita Kiribati 88
El Alamein Egypt 55
El Fasher Sudan 64
El Kharga Egypt 55
El Minya Egypt 55
El Obeid Sudan 64
El Paso United States 41
El Progreso Honduras 38
El Serrat Andorra 16
Elbasan Albania 16
Elblag Poland 26
Eldoret Kenya 57
Elmina Ghana 56
Emmen Netherlands 26
Encamp Andorra 16
Encarnación Paraguay 46
English Harbour Antigua & Barbuda 34
Enniskillen United Kingdom 31
Enschede Netherlands 26
Entebbe Uganda 66
Enugu Nigeria 62
Erdenet Mongolia 79
Erfurt Germany 21
Erzurum Turkey 29
Esbjerg Denmark 19
Esch-sur-Alzette Luxembourg 24
Eschen Liechtenstein 24
Escuintla Guatemala 38
Esfahan Iran 74
Esmeraldas Ecuador 45
Espoo Finland 21
Essen Germany 21
Ettelbrück Luxembourg 24
Évora Portugal 27
Exeter United Kingdom 30
Ezulwini Swaziland 65

F

Fada N'Gourma Burkina Faso 51
Fairbanks United States 41
Falealupo Western Samoa 91
Falevai Western Samoa 91
Falmouth Jamaica 39
Famagusta Cyprus 19

Fancy St Vincent 40
Farah Afghanistan 70
Faridpur Bangladesh 71
Farim Guinea Bissau 57
Faro Portugal 27
Faya-Largeau Chad 53
Fdérik Mauritania 60
Felixstowe United Kingdom 30
Fergana Uzbekistan 85
Ferkessédougou Côte D'Ivoire 54
Fez Morocco 61
Fianarantsoa Madagascar 59
Fier Albania 16
Fiorentino San Marino 23
Flatts Village Bermuda 35
Florence Italy 23
Flores Guatemala 38
Folkestone United Kingdom 30
Fontainebleau France 20
Fontvieille Monaco 20
Fort Shevchenko Kazakhstan 77
Fort Worth United States 41
Fortaleza Brazil 44
Foumban Cameroon 52
Francistown Botswana 51
Frankfurt-am-Main Germany 21
Fray Bentos Uruguay 47
Fredericton Canada 36
Frederikshåb Greenland 37
Frederikshavn Denmark 19
Freeport Bahamas 34
Freetown Sierra Leone 63
Fremantle Australia 88
Fria Guinea 57
Fuerte Olimpo Paraguay 46
Fuerteventura Canary Islands 52
Fujairah United Arab Emirates 85
Fukuoka Japan 75
Funafuti Tuvalu 88
Funchal Madeira 59
Fuzhou China 73

G

Gaalkacyo Somalia 64
Gabès Tunisia 66
Gaborone Botswana 51
Gabu Guinea Bissau 57
Gafsa Tunisia 66
Gagnoa Côte D'Ivoire 54
Gagra Georgia 72
Galap Palau 90
Galati Romania 27
Galle Sri Lanka 83
Galway Ireland 23
Gamboma Congo 54
Gander Canada 36
Ganta Liberia 58
Gao Mali 60
Garapan Northern Marianas 89
Garoua Cameroon 52
Gävle Sweden 28
Gaza Israel 75
Gaziantep Turkey 29
Gbadolite Zaïre 66
Gbarnga Liberia 58
Gdansk Poland 26
Gdynia Poland 26
Geelong Australia 88
Geneva Switzerland 29
Genoa Italy 23
George Town Malaysia 78
Georgetown St Vincent 40
Georgetown Guyana 46
Georgetown Gambia 56
Geraldton Australia 88
Germiston South Africa 64
Gevgelija Macedonia 25
Ghadamis Libya 58
Ghanzi Botswana 51
Ghat Libya 58
Ghazni Afghanistan 70
Ghent Belgium 17
Gibraltar Spain 28
Gijón Spain 28
Gisenye Rwanda 62
Gitega Burundi 52
Giurgiu Romania 27
Giza Egypt 55
Gjirokastër Albania 16
Glasgow United Kingdom 31
Goa India 74
Gobabis Namibia 61
Godhavn Greenland 37
Godthåb Greenland 37
Goiás Brazil 44
Goikul Palau 90
Gomel Belarus 17
Gonaïves Haiti 38
Gondar Ethiopia 55
Gospic Croatia 18
Göteborg Sweden 28
Gouyave Grenada 38
Governor's Harbour Bahamas 34
Granada Spain 28
Granada Nicaragua 39
Grand Bassam Côte D'Ivoire 54
Grand Roy Grenada 38
Grande Anse St Lucia 40
Grau-Roig Andorra 16
Graz Austria 16
Great Falls United States 41
Greenville Liberia 58
Greifswald Germany 21
Grenoble France 20
Grenville Grenada 38
Grodno Belarus 17
Groningen Netherlands 26
Grootfontein Namibia 61
Grøs Islets St Lucia 40
Guadalajara Mexico 39
Guanajuato Mexico 39
Guangzhou China 73
Guantanamo(US Base) Cuba 36
Guatemala City Guatemala 38
Guayaquil Ecuador 45
Gullin China 73
Gwadar Pakistan 80
Gweru Zimbabwe 67
Gyandzha Azerbaijan 70
Györ Hungary 22

H

Haapsalu Estonia 20
Haarlem Netherlands 26
Habban Yemen 85
Hadiboh Yemen 85
Haifa Israel 75
Haiphong Vietnam 85
Hajmah Oman 80
Hakodate Japan 75
Halab Syria 83
Halba Lebanon 78
Halifax Canada 36
Halle Germany 21
Hama Syria 83
Hamadan Iran 74
Hambantota Sri Lanka 83
Hamburg Germany 21
Hämeenlinna Finland 21
Hamhung North Korea 76
Hamilton Bermuda 35
Hamilton Canada 36
Hamilton New Zealand 89
Hammamet Tunisia 66
Hammerfest Norway 26
Hangzhou China 73
Hannover Germany 21
Hanoi Vietnam 85
Harar Ethiopia 55
Harare Zimbabwe 67
Harbin China 73
Hargeysa Somalia 64
Harper Liberia 58
Harwich United Kingdom 30
Hassi Messaoud Algeria 50
Hastings Barbados 34
Havana Cuba 36
Hebron Israel 75
Helsingborg Sweden 28
Helsingør Denmark 19
Helsinki Finland 21
Henzada Myanmar 79
Herakleion Greece 22
Herat Afghanistan 70
Heredia Costa Rica 35
Hermel Lebanon 78
Hillsborough Grenada 38
Hiroshima Japan 75
Hlatikulu Swaziland 65
Ho Ghana 56
Ho Chi Minh City Vietnam 85
Hobart Australia 88
Hobyo Somalia 64
Hodeida Yemen 85
Hoek van Holland Netherlands 26
Höfn Iceland 22
Hofuf Saudi Arabia 82
Holetown Barbados 34
Holguin Cuba 36
Holon Israel 75
Holyhead United Kingdom 31
Homs Syria 83
Hong Kong Hong Kong 73
Honiara Solomon Islands 90
Honolulu United States 41
Horta Azores 50
Houma Tonga 91
Houston United States 41
Hovd Mongolia 79
Hsinchu Taiwan 84
Hualien Taiwan 84
Huambo Angola 50
Hue Vietnam 85
Hull United Kingdom 30
Humacao Puerto Rico 40
Hungnam North Korea 76
Hurghada Egypt 55
Hwange Zimbabwe 67
Hyderabad India 74
Hyderabad Pakistan 80

I

Iasi Romania 27
Ibadan Nigeria 62
Ibiza Balearic Is. 17
Ibri Oman 80
Ica Peru 46
Ifni Morocco 61
Iguatu Brazil 44
Iloilo Philippines 81
Ilorin Nigeria 62
Impfondo Congo 54
In Salah Algeria 50
Inchon South Korea 76
Independence Belize 35
Indore India 74
Inhambane Mozambique 61
Innsbruck Austria 16
Interlaken Switzerland 29
Invercargill New Zealand 89
Inverness United Kingdom 31
Iquitos Peru 46
Iranshahr Iran 74
Irbid Jordan 76
Iringa Tanzania 65
Irkutsk Russia 82
Isafjördhur Iceland 22
Iskenderun Turkey 29
Islamabad Pakistan 80
Istanbul Turkey 29
Ita Paraguay 46
Izmir Turkey 29

J

Jackson United States 41
Jacmel Haiti 38
Jaffa Israel 75
Jaffna Sri Lanka 83
Jaipur India 74
Jakarta Indonesia 74
Jakobshavn Greenland 37
Jalalabad Afghanistan 70
Jarabulus Syria 83
Jayapura Indonesia 74
Jeddah Saudi Arabia 82
Jelgava Latvia 24
Jérémie Haiti 38
Jericho Israel 75
Jerusalem Israel 75
Jimma Ethiopia 55
Jinan China 73
Jinja Uganda 66
Joensuu Finland 21
Johannesburg South Africa 64
Johor Baharu Malaysia 78
Jönköping Sweden 28
Juba Sudan 64
Jubayl Saudi Arabia 82
Juneau United States 41
Jurmala Latvia 24
Jurong Singapore 83
Juticalpa Honduras 38
Jwaneng Botswana 51
Jyväskylä Finland 21

K

Kabala Sierra Leone 63
Kabale Uganda 66
Kabarole Uganda 66
Kabul Afghanistan 70
Kabwe Zambia 67
Kadoma Zimbabwe 67
Kaduna Nigeria 62
Kaédi Mauritania 60
Kaesong North Korea 76
Kagul Moldova 25
Kairouan Tunisia 66
Kalabera Northern Marianas 89
Kalai Mor Turkmenistan 84
Kalgoorlie Australia 88
Kalomo Zambia 67
Kalutara Sri Lanka 83
Kamo Armenia 70
Kampala Uganda 66
Kananga Zaïre 66
Kandahar Afghanistan 70
Kandy Sri Lanka 83
Kangnung South Korea 76
Kankan Guinea 57
Kano Nigeria 62
Kanpur India 74
Kansas City United States 41
Kanye Botswana 51
Kaohsiung Taiwan 84
Kaolack Senegal 63
Karachi Pakistan 80
Karaganda Kazakhstan 77
Karak Jordan 76
Karasburg Namibia 61
Karaudanawa Guyana 46
Karibib Namibia 61
Karlovac Croatia 18
Karlstad Sweden 28
Karonga Malawi 59
Kars Turkey 29
Kasama Zambia 67
Kasane Botswana 51
Kashi China 73
Kassala Sudan 64
Kassel Germany 21
Kasserine Tunisia 66
Kasungu Malawi 59
Kathmandu Nepal 80
Katong Singapore 83
Katowice Poland 26
Kaunas Lithuania 24
Kawasaki Japan 75
Kayes Mali 60
Kazanluk Bulgaria 18
Kecskemét Hungary 22
Kédougou Senegal 63
Keetmanshoop Namibia 61
Keflavik Iceland 22
Keklau Palau 90
Kenema Sierra Leone 63
Kenitra Morocco 61
Kenting Taiwan 84
Kerema Papua New Guinea 90
Kerewan Gambia 56
Kerki Turkmenistan 84
Kerman Iran 74
Khabarovsk Russia 82
Kharkov Ukraine 29
Khartoum Sudan 64
Khiva Uzbekistan 85
Khodzheni Tajikistan 84
Khon Kaen Thailand 84
Khor Fakkan United Arab Emirates 85
Khorog Tajikistan 84
Khulna Bangladesh 71
Khutse Botswana 51
Kibungu Rwanda 62
Kibuye Rwanda 62
Kicevo Macedonia 25
Kiel Germany 21
Kieta Papua New Guinea 90
Kiev Ukraine 29
Kigali Rwanda 62
Kigoma Tanzania 65
Kikwit Zaïre 66
Kilkenny Ireland 23
Killarney Ireland 23
Kilwe Kisiwani Tanzania 65
Kimberley South Africa 64
Kimchaek North Korea 76
Kindia Guinea 57
Kingston Jamaica 39
Kingstown St Vincent 40
Kinshasa Zaïre 66
Kipili Tanzania 65
Kirkuk Iraq 75
Kirovakan Armenia 70
Kiruna Sweden 28
Kisangani Zaïre 66
Kismaayo Somalia 64
Kisumu Kenya 57
Kitakyushu Japan 75
Kitale Kenya 57
Kitgum Uganda 66
Kitwe Zambia 67
Kizyl Arvat Turkmenistan 84
Klagenfurt Austria 16
Klaipeda Lithuania 24
Knock Ireland 23
Kobe Japan 75
Kocani Macedonia 25
Koforidua Ghana 56
Kohtla-Järve Estonia 20
Kokand Uzbekistan 85
Kokchetav Kazakhstan 77
Kolding Denmark 19
Köln Germany 21
Kolovai Tonga 91
Kompong Cham Cambodia 72
Kompong Chhnang Cambodia 72
Kompong Som Cambodia 72
Kompong Thom Cambodia 72
Komsomolsk-na-Amure Russia 82
Konrei Palau 90
Konya Turkey 29
Koper Slovenia 28
Korçë Albania 16
Korhogo Côte D'Ivoire 54
Koror Palau 90
Kos Greece 22
Kosice Slovak Republic 27
Kosti Sudan 64
Koszalin Poland 26
Kota Kinabalu Malaysia 78
Koudougou Burkina Faso 51
Kouroussa Guinea 57
Kowloon Hong Kong 73
Kozina Slovenia 28
Kpalimé Togo 65
Kraków Poland 26
Kranj Slovenia 28
Kranji Singapore 83
Krasnovodsk Turkmenistan 84
Krasnoyarsk Russia 82
Kribi Cameroon 52
Krishnanagar Bangladesh 71
Kristiansand Norway 26
Kristiansund Norway 26
Krujë Albania 16
Kuala Belait Brunei 72
Kuala Lumpur Malaysia 78
Kuala Terengganu Malaysia 78
Kuba Azerbaijan 70
Kuching Malaysia 78
Kukës Albania 16
Kulsary Kazakhstan 77
Kulyab Tajikistan 84
Kumanovo Macedonia 25
Kumasi Ghana 56
Kumayri Armenia 70
Kunfunadhoo Maldives 79
Kunming China 73
Kunsan South Korea 76
Kuopio Finland 21
Kuredhdhu Maldives 79
Kuressaare Estonia 20
Kurgan-Tyube Tajikistan 84
Kurunegala Sri Lanka 83
Kustanay Kazakhstan 77
Kutaisi Georgia 72
Kuwait City Kuwait 77
Kwangju South Korea 76
Kwekwe Zimbabwe 67
Kyongju South Korea 76
Kyoto Japan 75
Kyrenia Cyprus 19
Kzyl-Orda Kazakhstan 77

L

La Ceiba Honduras 38
La Condamine Monaco 20
La Paz Mexico 39
La Paz Bolivia 44
La Plaine Dominica 36
La Plata Argentina 44
La Romana Dominican Republic 37
La Serena Chile 45
La Union El Salvador 37
La Vega Dominican Republic 37
La'youne Western Sahara 61
Labasa Fiji 88
Labé Guinea 57
Laborie St Lucia 40
Lae Papua New Guinea 90
Laghouat Algeria 50
Lagos Portugal 27
Lagos Nigeria 62
Lahore Pakistan 80
Lahti Finland 21
Lalibela Ethiopia 55
Lalomanu Western Samoa 91
Lamap Vanuatu 91
Lambaréné Gabon 56
Lanzhou China 73
Lao Cai Vietnam 85
Laoag Philippines 81
Larisa Greece 22
Larnaca Cyprus 19
Larne United Kingdom 31
Las Palmas Canary Islands 52
Las Vegas United States 41
Lashio Myanmar 79
Lastoursville Gabon 56
Latakia Syria 83
Lausanne Switzerland 29
Lavumisa Swaziland 65
Layou St Vincent 40
Le Havre France 20
Leeds United Kingdom 30
Leeuwarden Netherlands 26
Lefka Cyprus 19
Legoland Denmark 19
Leiden Netherlands 26
Leipzig Germany 21
Lenkoran Azerbaijan 70
León Nicaragua 39
Leribe Lesotho 58
Les Allobroges Vanuatu 91
Les Cayes Haiti 38
Les Escaldes Andorra 16
Lesbos Greece 22
Leticia Colombia 45
Levuka Fiji 88
Lezhë Albania 16
Lhasa China 73
Libenge Zaïre 66
Liberec Czech Republic 19
Liberia Costa Rica 35
Libreville Gabon 56
Lichinga Mozambique 61
Liège Belgium 17
Liepaja Latvia 24
Likasi Zaïre 66
Lille France 20
Lilongwe Malawi 59
Lima Peru 46
Limassol Cyprus 19
Limbe Cameroon 52
Limbe Malawi 59
Limerick Ireland 23
Limoges France 20
Limón Costa Rica 35
Linden Guyana 46
Lindi Tanzania 65
Linguère Senegal 63
Linz Austria 16
Lisbon Portugal 27
Lisburn United Kingdom 31
Liverpool United Kingdom 30
Livno Bosnia-Herzegovina 18
Ljubljana Slovenia 28
Ljuw Nauru 89
Lobamba Swaziland 65
Lobatse Botswana 51
Lobito Angola 50
Lodwar Kenya 57
Lodz Poland 26
Loja Ecuador 45
Lomé Togo 65
Loméméti Vanuatu 91
London United Kingdom 30
London Kiribati 88
Longreach Australia 88
Los Angeles United States 41
Loubomo Congo 54
Luanda Angola 50
Luang-Phrabang Laos 78
Luba Equatorial Guinea 55
Lubango Angola 50
Lübeck Germany 21
Lublin Poland 26
Lubumbashi Zaïre 66
Lucknow India 74
Lüderitz Namibia 61
Luena Angola 50
Lugano Switzerland 29
Luganville Vanuatu 91
Lunsar Sierra Leone 63
Lusaka Zambia 67
Luuq Somalia 64
Luxembourg Luxembourg 24
Luxor Egypt 55
Luzern Switzerland 29
Lvov Ukraine 29
Lyon France 20

M

Ma'an Jordan 76
Maastricht Netherlands 26
Machakos Kenya 57
Machala Ecuador 45
Madang Papua New Guinea 90
Madhah Oman 80
Madras India 74
Madrid Spain 28
Mafeteng Lesotho 58
Mafraq Jordan 76
Magadan Russia 82
Magdeburg Germany 21
Mahajanga Madagascar 59
Mahébourg Mauritius 60
Mahón Balearic Is. 17
Maiduguri Nigeria 62
Makarska Croatia 18
Makeni Sierra Leone 63
Makokou Gabon 56
Malabo Equatorial Guinea 55
Málaga Spain 28
Malakal Sudan 64
Malanje Angola 50
Malaren Sweden 28
Malé Maldives 79
Malindi Kenya 57
Malmö Sweden 28
Man Côte D'Ivoire 54
Manacor Balearic Is. 17
Manado Indonesia 74
Managua Nicaragua 39
Manama Bahrain 71
Manaus Brazil 44
Manchester United Kingdom 30
Mandalay Myanmar 79
Mandeville Jamaica 39
Mango Togo 65
Mangot Dominica 36
Manila Philippines 81
Mankayane Swaziland 65
Manokwari Indonesia 74
Mansa Zambia 67
Mansoa Guinea Bissau 57
Manta Ecuador 45
Manzini Swaziland 65
Maputo Mozambique 61
Mar del Plata Argentina 44
Maracaibo Venezuela 47
Maradi Niger 62
Marakabeis Lesotho 58
Maramba Zambia 67
Maribor Slovenia 28
Mariental Namibia 61
Marigot St Lucia 40
Mariscal Estigarribia Paraguay 46
Marka Somalia 64
Maroua Cameroon 52

Marquis St Lucia 40
Marrakech Morocco 61
Marsabit Kenya 57
Marsaxlokk Malta 25
Marseille France 20
Mary Turkmenistan 84
Masaka Uganda 66
Masan South Korea 76
Masaya Nicaragua 39
Maseru Lesotho 58
Mashhad Iran 74
Masindi Uganda 66
Massawa Eritrea 55
Masuku Gabon 56
Masvingo Zimbabwe 67
Matadi Zaïre 66
Matagalpa Nicaragua 39
Matam Senegal 63
Matara Sri Lanka 83
Matrah Oman 80
Matrûh Egypt 55
Matthew Town Bahamas 34
Maun Botswana 51
Maura Brunei 72
Mauren Liechtenstein 24
Mavinga Angola 50
Mayagüez Puerto Rico 40
Mayumba Gabon 56
Mazar-i-Sharif Afghanistan 70
Mazatenango Guatemala 38
Mazeikiai Lithuania 24
Mbabane Swaziland 65
Mbala Zambia 67
Mbale Uganda 66
Mbandaka Zaïre 66
Mbé Congo 54
Mbeya Tanzania 65
Mbuji-Mayi Zaïre 66
Mdina Malta 25
Mecca Saudi Arabia 82
Mechelen Belgium 17
Medellín Colombia 45
Medenine Tunisia 66
Medina Saudi Arabia 82
Megri Armenia 70
Mejicanos El Salvador 37
Meknès Morocco 61
Melaka Malaysia 78
Melbourne Australia 88
Melilla Spain 28
Melo Uruguay 47
Memphis United States 41
Mendi Papua New Guinea 90
Mendoza Argentina 44
Menongue Angola 50
Mercedes Uruguay 47
Mergui Myanmar 79
Mérida Mexico 39
Mérida Venezuela 47
Mersin Turkey 29
Merthyr Tydfil United Kingdom 31
Mesters Vig Greenland 37
Mexicali Mexico 39
Mexico City Mexico 39
Miami United States 41
Micoud St Lucia 40
Milan Italy 23
Milford Haven United Kingdom 31
Milwaukee United States 41
Mina al Ahmadi Kuwait 77
Mina Saud Kuwait 77
Mina Sulman Bahrain 71
Minas Uruguay 47
Mindelo Cape Verde 53
Mingyan Myanmar 79
Minneapolis United States 41
Minsk Belarus 17
Miskolc Hungary 22
Misratah Libya 58
Mitu Colombia 45
Mitzic Gabon 56
Moçambique Mozambique 61
Mochudi Botswana 51
Moengo Surinam 47
Mogadishu Somalia 64
Mogilev Belarus 17
Mohales Hoek Lesotho 58
Mokha Yemen 85
Mokhotlong Lesotho 58
Mokolo Cameroon 52
Mokpo South Korea 76
Moldovita Romania 27
Mollendo Peru 46
Mombasa Kenya 57
Monaco Monaco 20
Monastir Tunisia 66
Mongo Chad 53
Monkey Bay Malawi 59
Monrovia Liberia 58
Mons Belgium 17
Monte Carlo Monaco 20
Montego Bay Jamaica 39
Monterrey Mexico 39
Montevideo Uruguay 47
Montréal Canada 36
Montreux Switzerland 29
Mopti Mali 60
Morogoro Tanzania 65
Morondava Madagascar 59
Moroni Comoros 54
Moroto Uganda 66
Moruga Trinidad & Tobago 41
Moscow Russia 82
Moshi Tanzania 65
Mosta Malta 25
Mostaganem Algeria 50
Mostar Bosnia-Herzegovina 18
Mosul Iraq 75
Moulmein Myanmar 79
Moundou Chad 53
Mount Darwin Zimbabwe 67
Mount Hagen Papua New Guinea 90
Mount Isa Australia 88
Moyale Kenya 57
Mozyr Belarus 17
Mtwara Tanzania 65
Mu'a Tonga 91
Mufulira Zambia 67
Muharraq Bahrain 71
Mukalla Yemen 85
Mukeru Palau 90
Mullaittivu Sri Lanka 83
Multan Pakistan 80
München Germany 21
Münster Germany 21
Muramvya Burundi 52
Murcia Spain 28
Murgab Tajikistan 84
Murmansk Russia 82
Muscat Oman 80
Mutare Zimbabwe 67
Muyinga Burundi 52
Mwanza Tanzania 65
Myitkyina Myanmar 79
Mymensingh Bangladesh 71
Mzuzu Malawi 59

N

N'Djaména Chad 53
N'kayi Congo 54
Nablus Israel 75
Nacala Mozambique 61
Nadi Fiji 88
Nagasaki Japan 75
Nagoya Japan 75
Nagpur India 74
Nairobi Kenya 57
Najran Saudi Arabia 82
Nakhichevan Azerbaijan 70
Nakhon Ratchasima Thailand 84
Nakhon Sawan Thailand 84
Nakhon Si Thammarat Thailand 84
Nakuru Kenya 57
Namangan Uzbekistan 85
Namibe Angola 50
Nampo North Korea 76
Nampula Mozambique 61
Namur Belgium 17
Nanchang China 73
Nancy France 20
Nanjing China 73
Nanning China 73
Nantes France 20
Napier New Zealand 89
Naples Italy 23
Narayanganj Bangladesh 71
Narsarsuaq Greenland 37
Narva Estonia 20
Narvik Norway 26
Naryn Kyrgyzstan 77
Nashville United States 41
Nassau Bahamas 34
Natitingou Benin 51
Nausori Fiji 88
Nazareth Israel 75
Nazca Peru 46
Nazwa Oman 80
Ndjole Gabon 56
Ndola Zambia 67
Nebit-Dag Turkmenistan 84
Negombo Sri Lanka 83
Negril Jamaica 39
Neiafu Tonga 91
Nelson New Zealand 89
Néma Mauritania 60
Nendeln Liechtenstein 24
Neuchâtel Switzerland 29
New Amsterdam Guyana 46
New Delhi India 74
New Orleans United States 41
New Sandy Bay St Vincent 40
New Tamale Ghana 56
New York United States 41
Newcastle United Kingdom 30
Newcastle St Christopher/Nevis 40
Newcastle Australia 88
Newhaven United Kingdom 30
Newport United Kingdom 31
Newry United Kingdom 31
Ngaoundéré Cameroon 52
Ngarekeuk Palau 90
Ngatpang Palau 90
Ngozi Burundi 52
Nha Trang Vietnam 85
Nhlangano Swaziland 65
Niamey Niger 62
Nice France 20
Nicosia Cyprus 19
Nicoya Costa Rica 35
Niefang Equatorial Guinea 55
Nieuw Nickerie Surinam 47
Nikolayev Ukraine 29
Nis Yugoslavia 29
Nisab Yemen 85
Nitra Slovak Republic 27
Nizhny Novgorod Russia 82
Nkongsamba Cameroon 52
Norrköping Sweden 28
Norsoup Vanuatu 91
Norwich United Kingdom 30
Nouadhibou Mauritania 60
Nouakchott Mauritania 60
Novabad Tajikistan 84
Novosibirsk Russia 82
Nuku'alofa Tonga 91
Nukus Uzbekistan 85
Nürnberg Germany 21
Nushki Pakistan 80
Nuwara Eliya Sri Lanka 83
Nzérékoré Guinea 57

O

Oaxaca Mexico 39
Oban United Kingdom 31
Obock Djibouti 54
Obuasi Ghana 56
Ocho Rios Jamaica 39
Odense Denmark 19
Odessa Ukraine 29
Odienné Côte D'Ivoire 54
Ogbomosho Nigeria 62
Ohrid Macedonia 25
Okha Russia 82
Okhotsk Russia 82
Old Road St Christopher/Nevis 40
Olinda Brazil 44
Olomouc Czech Republic 19
Omdurman Sudan 64
Omsk Russia 82
Oporto Portugal 27
Oradea Romania 27
Oran Algeria 50
Orange Walk Belize 35
Oranjemund Namibia 61
Orapa Botswana 51
Ordino Andorra 16
Örebro Sweden 28
Orgeyev Moldova 25
Orlando United States 41
Orsha Belarus 17
Oruro Bolivia 44
Osaka Japan 75
Osh Kyrgyzstan 77
Oshogbo Nigeria 62
Osijek Croatia 18
Oslo Norway 26
Ostend Belgium 17
Östersund Sweden 28
Ostrava Czech Republic 19
Oswiecim Poland 26
Ottawa Canada 36
Ouagadougou Burkina Faso 51
Ouahigouya Burkina Faso 51
Ouargla Algeria 50
Ouesso Congo 54
Ouidah Benin 51
Oulu Finland 21
Owando Congo 54
Oxford United Kingdom 30
Oyo Nigeria 62

P

Padang Indonesia 74
Pagan Myanmar 79
Pakse Laos 78
Pal Andorra 16
Palapye Botswana 51
Palembang Indonesia 74
Palermo Italy 23
Palma Balearic Is. 17
Palmerston North New Zealand 89
Panama City Panama 39
Panevezys Lithuania 24
Pang-Pang Vanuatu 91
Pangai Tonga 91
Paphos Cyprus 19
Paraguari Paraguay 46
Parakou Benin 51
Paramaribo Surinam 47
Pardubice Czech Republic 19
Paris France 20
Paris Kiribati 88
Pärnu Estonia 20
Paro Bhutan 71
Pas de la Casa Andorra 16
Pasto Colombia 45
Patan Nepal 80
Patras Greece 22
Pattaya Thailand 84
Paviodar Kazakhstan 77
Paysandú Uruguay 47
Pécs Hungary 22
Pedra Lume Cape Verde 53
Pegu Myanmar 79
Pemba Mozambique 61
Perth United Kingdom 31
Perth Australia 88
Pescara Italy 23
Peshawar Pakistan 80
Petaling Jaya Malaysia 78
Petit Loango Gabon 56
Petropavlovsk Kazakhstan 77
Philadelphia United States 41
Phitsanulok Thailand 84
Phnom Penh Cambodia 72
Phoenix United States 41
Phuket Thailand 84
Phuntsholing Bhutan 71
Pietermaritzburg South Africa 64
Pigg's Peak Swaziland 65
Pilar Paraguay 46
Pinsk Belarus 17
Piraeus Greece 22
Piran Slovenia 28
Pisa Italy 23
Piteå Sweden 28
Pittsburgh United States 41
Pleven Bulgaria 18
Plitvice Croatia 18
Ploiesti Romania 27
Plovdiv Bulgaria 18
Plymouth United Kingdom 30
Plzen Czech Republic 19
Podgorica Yugoslavia 29
Point Fortin Trinidad & Tobago 41
Pointe-Noire Congo 54
Pokhara Nepal 80
Polonnaruwa Sri Lanka 83
Polotsk Belarus 17
Poltava Ukraine 29
Ponce Puerto Rico 40
Ponta Delgada Azores 50
Pontianak Indonesia 74
Pontoetoe Surinam 47
Pori Finland 21
Port Antonio Jamaica 39
Port Augusta Australia 88
Port Elizabeth St Vincent 40
Port Elizabeth South Africa 64
Port Gentil Gabon 56
Port Harcourt Nigeria 62
Port Kaituma Guyana 46
Port Louis Mauritius 60
Port Moresby Papua New Guinea 90
Port of Spain Trinidad & Tobago 41
Port Said Egypt 55
Port Sudan Sudan 64
Port-au-Prince Haiti 38
Port-de-Paix Haiti 38
Port-Vila Vanuatu 91
Portalegre Portugal 27
Portland United States 41
Pôrto Alegre Brazil 44
Porto do Moniz Madeira 59
Porto Novo Benin 51
Porto Santo Madeira 59
Porto-Alegre São Tomé & Principe 63
Portobelo Panama 39
Portoviejo Ecuador 45
Portsmouth United Kingdom 30
Portsmouth Dominica 36
Poti Georgia 72
Potnarhvin Vanuatu 91
Potosí Bolivia 44
Pottuvil Sri Lanka 83
Poznan Poland 26
Praia Cape Verde 53
Praslin St Lucia 40
Presov Slovak Republic 27
Pretoria South Africa 64
Prilep Macedonia 25
Przhevalsk Kyrgyzstan 77
Ptuj Slovenia 28
Puebla Mexico 39
Puerto Barrios Guatemala 38
Puerto Cabezas Nicaragua 39
Puerto Carreño Colombia 45
Puerto Gortés Honduras 38
Puerto la Cruz Venezuela 47
Puerto Lempira Honduras 38
Puerto Plata Dominican Republic 37
Pujehun Sierra Leone 63
Pula Croatia 18
Punakha Bhutan 71
Pune India 74
Punta Arenas Chile 45
Punta del Este Uruguay 47
Punta Gorda Belize 35
Puntarenas Costa Rica 35
Pursat Cambodia 72
Pusan South Korea 76
Puttalam Sri Lanka 83
Pyongyang North Korea 76

Q

Qachas Nek Lesotho 58
Qardho Somalia 64
Qingdao China 73
Qom Iran 74
Qormi Malta 25
Québec Canada 36
Queenstown New Zealand 89
Quelimane Mozambique 61
Quetta Pakistan 80
Quezaltenango Guatemala 38
Quezon City Philippines 81
Qui Nhon Vietnam 85
Quito Ecuador 45
Quthing Lesotho 58

R

Rabat Malta 25
Rabat Morocco 61
Rabat(Gozo) Malta 25
Rabaul Papua New Guinea 90
Radom Poland 26
Ramsgate United Kingdom 30
Randers Denmark 19
Rangoon Myanmar 79
Ras ad Daqm Oman 80
Ras al Khaimah U.A.E. 85
Ras an Naqab Jordan 76
Ras Laffan Qatar 81
Rasht Iran 74
Rawalpindi Pakistan 80
Recife Brazil 44
Redange Luxembourg 24
Reggio di Calabria Italy 23
Regina Canada 36
Reno United States 41
Reykjavík Iceland 22
Rezekne Latvia 24
Rhodes Greece 22
Riaba Equatorial Guinea 55
Riberalta Bolivia 44
Richmond United States 41
Riga Latvia 24
Rijeka Croatia 18
Rila Bulgaria 18
Rincón Costa Rica 35
Rio Branco Brazil 44
Rio de Janeiro Brazil 44
Riobamba Ecuador 45
Rivera Uruguay 47
Rivière du Rampart Mauritius 60
Robertsport Liberia 58
Roboré Bolivia 44
Rocha Uruguay 47
Rockhampton Australia 88
Rome Italy 23
Ronave Nauru 89
Røros Norway 26
Rosalie Dominica 36
Rosario Argentina 44
Rose Hill Mauritius 60
Roseau Dominica 36
Rosslare Ireland 23
Rosso Mauritania 60
Rostock Germany 21
Rostov-on-Don Russia 82
Rotterdam Netherlands 26
Rovaniemi Finland 21
Rovno Ukraine 29
Roxborough Trinidad & Tobago 41
Rubona Rwanda 62
Ruhengeri Rwanda 62
Ruse Bulgaria 18
Rustavi Georgia 72
Rutana Burundi 52
Ruwais Qatar 81
Ruwais United Arab Emirates 85
Ruwayshid Jordan 76
Ruyigi Burundi 52
Rybnitsa Moldova 25

S

Saarbrücken Germany 21
Sabha Libya 58
Sabiyah Kuwait 77
Sad'ah Yemen 85
Sadao Thailand 84
Saint Joseph Dominica 36
Saipan Palau 90
Salalah Oman 80
Salamanca Spain 28
Salima Malawi 59
Salisbury Dominica 36
Salt Jordan 76
Salt Lake City United States 41
Salto Uruguay 47
Salvador Brazil 44
Salwah Qatar 81
Salyany Azerbaijan 70
Salzburg Austria 16
Samara Russia 82
Samarkand Uzbekistan 85
Samsun Turkey 29
San Antonio Belize 35
San Antonio United States 41
San Carlos de Bariloche Argentina 44
San Cristóbal Venezuela 47
San Diego United States 41
San Fernando Trinidad & Tobago 41
San Fernando de Apure Venezuela 47
San Francisco United States 41
San Fran. de Macorís Dom. Republic 37
San Ignacio Belize 35
San Jose United States 41
San José Costa Rica 35
San José Guatemala 38
San José Uruguay 47
San Juan Dominican Republic 37
San Juan Puerto Rico 40
San Marino San Marino 23
San Miguel El Salvador 37
San Pedro de Macorís Dom. Rep. 37
San Pedro Sula Honduras 38
San Salvador El Salvador 37
San Vicente El Salvador 37
Sana Yemen 85
Sandakan Malaysia 78
Sandy Point St Christopher/Nevis 40
Sangre Grande Trinidad & Tobago 41
Sanniquellie Liberia 58
Sans-Souci Haiti 38
Sant-Julià-de-Lòria Andorra 16
Santa Ana El Salvador 37
Santa Clara Cuba 36
Santa Cruz Bolivia 44
Santa Cruz Azores 50
Santa Cruz Canary Islands 52
Santa Cruz Madeira 59
Santa Marta Colombia 45
Santana Madeira 59
Santander Spain 28
Santarém Portugal 27
Santarém Brazil 44
Santiago Dominican Republic 37
Santiago Panama 39
Santiago Chile 45
Santiago de Compostela Spain 28
Santiago de Cuba Cuba 36
Säntis Switzerland 29
Santo Domingo Dominican Rep. 37
São Luís Brazil 44
São Paulo Brazil 44
São Tomé São Tomé and Principe 63
Sapporo Japan 75
Sarajevo Bosnia-Herzegovina 18
Sarandë Albania 16
Sarh Chad 53
Sarmiento Argentina 44
Saskatoon Canada 36
Sassandra Côte D'Ivoire 54
Sauteurs Grenada 38
Savalou Benin 51
Savannakhet Laos 78
Savé Benin 51
Sayhut Yemen 85
Scarborough United Kingdom 30
Scarborough Trinidad & Tobago 41
Schaan Liechtenstein 24
Scoresbysund Greenland 37
Seattle United States 41
Sefadu Sierra Leone 63
Ségou Mali 60
Sekondi Ghana 56
Selebi-Phikwe Botswana 51
Seletar Singapore 83
Semey Kazakhstan 77
Sendai Japan 75
Seoul South Korea 76
Serangoon Singapore 83
Seria Brunei 72
Serravalle San Marino 23
Serrekunda Gambia 56
Setúbal Portugal 27
Sevastopol Ukraine 29
Seville Spain 28
Sfax Tunisia 66
Sha Tin Hong Kong 73
Shanghai China 73
Sharjah United Arab Emirates 85
Sheffield United Kingdom 30
Shek Wu Hui Hong Kong 73
Shenyang China 73
Shenzhen China 73
Shibam Yemen 85
Shiraz Iran 74
Shkodër Albania 16
Shuaiba Kuwait 77
Shuwaikh Kuwait 77
Siauliai Lithuania 24
Sibenik Croatia 18
Sibiti Congo 54
Sidon Lebanon 78
Siem Reap Cambodia 72
Sigiriya Sri Lanka 83
Siguiri Guinea 57

Sikasso Mali 60
Silgarhi Nepal 80
Simferopol Ukraine 29
Simla India 74
Sines Portugal 27
Singapore City Singapore 83
Sinop Turkey 29
Sinuiju North Korea 76
Sisophon Cambodia 72
Siteki Swaziland 65
Sittwe Myanmar 79
Sivas Turkey 29
Siwa Egypt 55
Skikda Algeria 50
Skocjan Slovenia 28
Skopje Macedonia 25
Sligo Ireland 23
Sodankyla Finland 21
Sofia Bulgaria 18
Sohar Oman 80
Sokodé Togo 65
Sokoto Nigeria 62
Soldeu Andorra 16
Solwezi Zambia 67
Somerset Bermuda 35
Søndre Strømfjord Greenland 37
Songkhla Thailand 84
Sonsonate El Salvador 37
Sopron Hungary 22
Soroki Moldova 25
Soroti Uganda 66
Souanké Congo 54
Soufrière St Lucia 40
Souillac Mauritius 60
Sousse Tunisia 66
Southampton United Kingdom 30
Spanish Town Jamaica 39
Speightstown Barbados 34
Spitak Armenia 70
Split Croatia 18
Sri-Jayawardenapura Sri Lanka 83
Srinagar India 74
St Gallen Switzerland 29
St George Bermuda 35
St Georges Grenada 38
St John's Antigua & Barbuda 34
St John's Canada 36
St Louis United States 41
St Moritz Switzerland 29
St Paul United States 41
St Paul's St Christopher/Nevis 40
St Paul's Bay Malta 25
St. David's United Kingdom 31
St. Louis Senegal 63
St. Petersburg Russia 82
Stanley Hong Kong 73
Stara Zagora Bulgaria 18
Stavanger Norway 26
Stepanakert Azerbaijan 70
Stip Macedonia 25
Sto. António São Tomé and Príncipe 63
Stockholm Sweden 28
Strabane United Kingdom 31
Stranraer United Kingdom 31
Strasbourg France 20
Strumica Macedonia 25
Stung Treng Cambodia 72
Stuttgart Germany 21
Sucre Bolivia 44
Suez Egypt 55
Sukhumi Georgia 72
Sukkur Pakistan 80
Sumgait Azerbaijan 70
Sundsvall Sweden 28
Sunyani Ghana 56
Sur Oman 80
Surabaya Indonesia 74
Surat Thani Thailand 84
Suva Fiji 88
Suzhou China 73
Swakopmund Namibia 61
Swansea United Kingdom 31
Sydney Australia 88
Szczecin Poland 26
Szeged Hungary 22
Székesfehérvár Hungary 22

T

Ta'izz Yemen 85
Tabeng Cambodia 72
Tabernacle St Christopher/Nevis 40
Tabora Tanzania 65
Tabriz Iran 74
Tachungnya Northern Marianas 89
Tadjoura Djibouti 54
Tadmur Syria 83
Taegu South Korea 76
Taejon South Korea 76
Tahoua Niger 62
Tai Po Hong Kong 73
Taichung Taiwan 84
Taif Saudi Arabia 82
Tainan Taiwan 84
Taipei Taiwan 84
Taitung Taiwan 84
Takamatsu Japan 75
Takeo Cambodia 72
Takoradi Ghana 56
Talara Peru 46
Talcahuano Chile 45
Taldy-Kurgan Kazakhstan 77
Tallinn Estonia 20
Tamale Ghana 56
Tamanrasset Algeria 50
Tambacounda Senegal 63
Tambora Indonesia 74
Tampa United States 41
Tampere Finland 21
Tampico Mexico 39
Tanapag Northern Marianas 89
Tanga Tanzania 65
Tangier Morocco 61
Tanout Niger 62
Taranto Italy 23
Tarija Bolivia 44
Tarim Yemen 85
Tartu Estonia 20
Tartus Syria 83
Tashauz Turkmenistan 84
Tashigang Bhutan 71
Tashkent Uzbekistan 85
Tbilisi Georgia 72
Tegucigalpa Honduras 38
Tehran Iran 74
Tel Aviv Israel 75
Telavi Georgia 72
Tema Ghana 56
Tenkodogo Burkina Faso 51
Tete Mozambique 61
Tétouan Morocco 61
Tetovo Macedonia 25
Teyateyaneng Lesotho 58
Thakhek Laos 78
The Crane Barbados 34
The Hague Netherlands 26
Thessaloniki Greece 22
Thiès Senegal 63
Thimphu Bhutan 71
Thule Greenland 37
Tianjin China 73
Tibati Cameroon 52
Tidjikdja Mauritania 60
Tiemcen Algeria 50
Tijuana Mexico 39
Timaru New Zealand 89
Timbuktu Mali 60
Timisoara Romania 27
Tindouf Algeria 50
Tiranë Albania 16
Tiraspol Moldova 25
Toalanaro Madagascar 59
Toamasina Madagascar 59
Tobruk Libya 58
Toco Trinidad & Tobago 41
Tokyo Japan 75
Toledo Spain 28
Toliara Madagascar 59
Tongsa Bhutan 71
Tornio Finland 21
Toronto Canada 36
Tororo Uganda 66
Tortosa Spain 28
Torún Poland 26
Toulouse France 20
Tournai Belgium 17
Tours France 20
Townsville Australia 88
Tozeur Tunisia 66
Trabzon Turkey 29
Tralee Ireland 23
Travnik Bosnia-Herzegovina 18
Trier Germany 21
Triesen Liechtenstein 24
Trieste Italy 23
Trincomalee Sri Lanka 83
Trinidad Cuba 36
Trinidad Bolivia 44
Tripoli Libya 58
Tripoli Lebanon 78
Tromsø Norway 26
Trondheim Norway 26
Trujillo Honduras 38
Trujillo Peru 46
Tsau Botswana 51
Tsévié Togo 65
Tshabong Botswana 51
Tshane Botswana 51
Tsuen Wan Hong Kong 73
Tsumeb Namibia 61
Tuas Singapore 83
Tuasivi Western Samoa 91
Tucker's Town Bermuda 35
Tucumán Argentina 44
Tuen Mun Hong Kong 73
Tulcán Ecuador 45
Tumaco Colombia 45
Tumbes Peru 46
Tunis Tunisia 66
Turin Italy 23
Turku Finland 21
Turrialba Costa Rica 35
Tutong Brunei 72
Tuzla Bosnia-Herzegovina 18
Tyre Lebanon 78

U

Uaboe Nauru 89
Ubon Ratchathani Thailand 84
Udon Thani Thailand 84
Uige Angola 50
Ujung Pandang Indonesia 74
Ulan Bator Mongolia 79
Ullapool United Kingdom 31
Umeå Sweden 28
Umm al Qaiwain U.A.E. 85
Umm Bab Qatar 81
Umm Qasr Kuwait 77
Umm Sai'd Qatar 81
Uppsala Sweden 28
Ura-Tyube Tajikistan 84
Uralsk Kazakhstan 77
Urumqi China 73
Ushuaia Argentina 44
Uskemen Kazakhstan 77
Utrecht Netherlands 26

V

Vaasa Finland 21
Vacoas Mauritius 60
Vaduz Liechtenstein 24
Valdivia Chile 45
Valencia Spain 28
Valencia Venezuela 47
Valladolid Spain 28
Valletta Malta 25
Valmiera Latvia 24
Valparaíso Chile 45
Vancouver Canada 36
Vantaa Finland 21
Varazdin Croatia 18
Varna Bulgaria 18
Vasteras Sweden 28
Vatican City Italy 23
Vatomandry Madagascar 59
Växjö Sweden 28
Veles Macedonia 25
Venice Italy 23
Ventspils Latvia 24
Veracruz Mexico 39
Viborg Denmark 19
Victoria Canada 36
Victoria Grenada 38
Victoria Seychelles 63
Vieng Sai Laos 78
Vienna Austria 16
Vientiane Laos 78
Vieux Fort St Lucia 40
Vigo Spain 28
Vila do Porto Azores 50
Villa Montes Bolivia 44
Villarrica Paraguay 46
Vilnius Lithuania 24
Viña del mar Chile 45
Vinh Vietnam 85
Virovitica Croatia 18
Vitebsk Belarus 17
Vladivostok Russia 82
Vlissingen Netherlands 26
Vlorë Albania 16
Vogan Togo 65
Voi Kenya 57
Voinjama Liberia 58
Volgograd Russia 82
Volos Greece 22
Vorkuta Russia 82
Vukovar Croatia 18

W

Wad Medani Sudan 64
Wadi Haifa Sudan 64
Wajir Kenya 57
Wallibu St Vincent 40
Walvis Bay Namibia 61
Wanganui New Zealand 89
Warsaw Poland 26
Washington United States 41
Waterford Ireland 23
Wau Sudan 64
Wellington New Zealand 89
Welshpool United Kingdom 31
Wesley Dominica 36
Weymouth United Kingdom 30
Whangarei New Zealand 89
Whitehorse Canada 36
Wick United Kingdom 31
Wiltz Luxembourg 24
Windhoek Namibia 61
Winnipeg Canada 36
Winterthur Switzerland 29
Wonsan North Korea 76
Woodlands Singapore 83
Worthing Barbados 34
Wrexham United Kingdom 31
Wroclaw Poland 26
Wuhan China 73

X

Xaafuun Somalia 64
Xai-Xai Mozambique 61
Xi'an China 73
Xieng Khouang Laos 78

Y

Yakutsk Russia 82
Yamoussoukro Côte D'Ivoire 54
Yangambi Zaïre 66
Yangor Nauru 89
Yaoundé Cameroon 52
Yaren Nauru 89
Yekaterinburg Russia 82
Yellowknife Canada 36
Yerevan Armenia 70
Yeviakh Azerbaijan 70
Yirga 'Alem Ethiopia 55
Yogyakarta Indonesia 74
Yokadouma Cameroon 52
Yokohama Japan 75
Yola Nigeria 62
York United Kingdom 30
Yuen Long Hong Kong 73

Z

Zadar Croatia 18
Zagreb Croatia 18
Zahle Lebanon 78
Zakataly Azerbaijan 70
Zakopane Poland 26
Zambezi Zambia 67
Zamboanga Philippines 81
Zanzibar Tanzania 65
Zaporozhye Ukraine 29
Zaragoza Spain 28
Zarqa Jordan 76
Zeebrugge Belgium 17
Zémio Central African Republic 53
Zenica Bosnia-Herzegovina 18
Ziguinchor Senegal 63
Zilina Slovak Republic 27
Zinder Niger 62
Zion St Christopher/Nevis 40
Zomba Malawi 59
Zouîrât Mauritania 60
Zürich Switzerland 29
Zvishavane Zimbabwe 67
Zwedru Liberia 58
Zwolle Netherlands 26